THE EMPIRE OF
NON-SENSE

THE EMPIRE OF NON-SENSE

ART IN THE TECHNOLOGICAL SOCIETY

JACQUES ELLUL

Introductory essays by Samir Younés & David Lovekin

Translated by Michael Johnson & David Lovekin

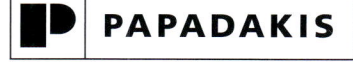

English edition first published in 2014 by Papadakis Publisher

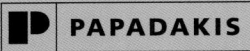

An imprint of New Architecture Group Limited

Kimber Studio, Winterbourne, Berkshire, RG20 8AN, UK
info@papadakis.net | www.papadakis.net

 @papadakisbooks   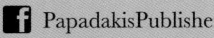 PapadakisPublisher

Originally published as *L'empire du non-sens* © Presses Universitaires de France 1980

Edited by Samir Younés

Publishing Director: Alexandra Papadakis
Design Director: Aldo Sampieri
House Editor: Sheila de Vallée
Production: Caroline Kuhtz

ISBN 978 1 906506 40 7

Copyright © 2014 Papadakis Publisher

Introductory essays copyright © 2014 Samir Younés and David Lovekin
Translation from the French copyright © 2014 Michael Johnson and David Lovekin
All rights reserved

Samir Younés and David Lovekin hereby assert their moral right to be identified as the authors of this work.
Michael Johnson and David Lovekin hereby assert their moral right to be identified as the translators of this work.

No part of this publication may be reproduced or transmitted in any form or by any means, electronic or mechanical, including photocopy, recording or any other information storage and retrieval system, without prior permission in writing from the Publisher.

A CIP catalogue record of this book is available from the British Library

Printed and bound in China

Acknowledgments
Samir Younés, David Lovekin, and Michael Johnson wish to thank the University of Notre Dame's School of Architecture for its generous support for the publication of this book.
Samir Younés, Editor

Cover photograph: David Lovekin

CONTENTS

Jacques Ellul and the Eclipse of Artistic Symbolism   7
Samir Younés

Art and Technology – All or Nothing   21
David Lovekin

THE EMPIRE OF NON-SENSE

Introduction   35

Chapter 1: **The Contradiction**   45

Chapter 2: **Art in the Technical System**   63

Chapter 3: **The Message and its Compensation**   85

Chapter 4: **Formalism and Theory**   111

Chapter 5: **The Artist and the Critic**   141

Afterword: **The Indefinite Future**   161

Biographical Notes   168

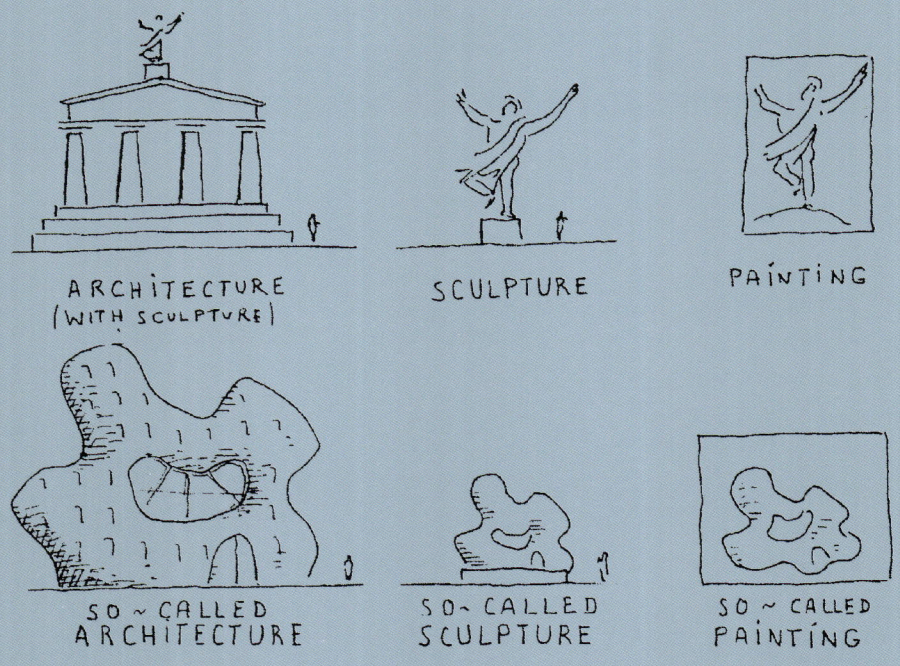

Fig. 1: Léon Krier, *Architecture, Sculpture, Painting*

"Art has become one of the major functions used to integrate humankind into the technical complex."

Jacques Ellul

JACQUES ELLUL
and the Eclipse of Artistic Symbolism

Samir Younés

Advocates of modernism have acclaimed it as an enlightened cultural force that reshaped a previously traditional society into a new and progressive society based on the liberating impetus of technology. Artists and architects who were protagonists of artistic modernism explained it as an artistic ideology that derived from a technological world-view while they actively worked at establishing the visual and verbal forms that should characterize this world-view. Following their broad diffusion on a planetary scale, these visual and verbal forms came to decisively qualify modern society as a technological society. Modernism was the only form of modernity that is suitable for the technological society, and the visual and verbal forms of the technological age were its inevitable expressions. History, in the broadest sense of the march of cultural forces, was qualified by an implacable arrow that teleologically led to the cultural apotheosis of modernism. Visual and verbal forms that did not espouse modernism were at best anachronistic, or worse, regressive delusions. These were some of the core beliefs of the official narrative of modernism as expressed in professional artistic production and in academic circles since the early decades of the twentieth century. In establishing this narrative, artists, architects, and historians linked their arguments to issues concerning the philosophy of history, of time, of cultural becoming and historical determinism, and the technical means by which artistic making is realized. Artistic and architectural modernism were explained as having risen as if by historical necessity. An explanation, however, no matter how relentlessly repeated, is not necessarily a justification. To say that massive technological means and products radically transformed art and architecture is a descriptive explanation, not necessarily a sound justification for why art and architecture should have been transformed in modernist ways. To say that there have been massive changes in means does not necessarily justify changing the ends of visual and verbal forms.

Here is where Jacques Ellul's (1912-1994) thought enters in full force. As one of the sharpest critics of the technological society Ellul clarified that technology was not a neutral set of means that serves human will while increasing human freedom. Indeed, the visual and verbal forms of the technological society structurally mold the mind in a technological direction while excluding other directions. If multiple traditional societies were symbolized by their visual and verbal forms, can it likewise be said that the technological society is symbolized by its own visual and verbal forms in the same way? Does the artistic symbolic function (e.g. artistic expression and representation) operate with equal ease within traditional societies and the technological society? Is the technological order itself symbolizable? Or has symbolic thought been radically impoverished; has it become an empire of non-sense as Ellul contends in this book? Before engaging this question, we might briefly consider symbolic thought within the visual and verbal cultures prior to modernism and the technological society.

The poietic order

Much of our intellectual evolution unfolds between our roles as perceivers and makers of visual and verbal culture. We receive visual and verbal cultures either unquestionably by virtue of being born within a certain culture, or judiciously through the intellectual choices we make; and we project onto them meanings of a private nature (personal taste, individual artistic judgment) as well as of a social nature (cultural taste, collective artistic judgment). This is, broadly speaking, the point of view of those who observe, desire, and use visual and verbal forms. For a more comprehensive view we need to account for the makers, those who actively construct visual and verbal forms and give them to society. We gain a deeper understanding of the image (painting, sculpture, architecture) and the word (prose, poetry, philosophy) when we consider them also from the standpoint of their maker, or rather their *poietic* maker—as in *poiesis*, from the Greek verb *poein*, to make, sculpt, fabricate. Making with images and making with words are two ways to construct a part of the world based on the partial perspective on the world available to painters or architects, to writers or poets. Taken together, these partial constructions of the world construct a certain society and, in turn, the relationship between this society and other societies, and the place they occupy within nature. Through repeated reciprocal influences the act of constructing and the act of understanding form and conform each other at once symbolically and literally. Constructing the world symbolically and literally is a way to understand the world symbolically and literally. Meaning arises from the mind reflecting upon itself and upon its own processes in relation to its own visual and verbal constructions. Meaning arises from the symbolic relation between the visual and verbal forms in a given context as well as the transcendence of this very context.

As in metaphor, symbolic thought correlates an initial meaning to a second meaning both visually and verbally. Symbolic thought allows for an understanding of the world that is at once rooted in the factual while referring to the true, realizing the particular while signifying the universal, based on the contingent while denoting the enduring, embedded in the immanent while symbolizing the transcendent. *In the poietic order then, artistic making can be said to have a purpose other than itself, that is, a purpose other than its own concreteness as a physical reality.* Putting it differently, visual works and verbal works gain their meaning by virtue of their links between the factual, contingent, and immanent, in relation to the enduring, universal, and transcendent. This is why for millennia artistic making cultivated its multiple expressions within a triangular relationship: 1) the relation between art and nature, 2) art and truth, and 3) art and reality. Art and nature concerns Nature understood in her laws and nature understood in her products (e.g. the proportions of the human body within the membrature of that body). Artistic truth pertains to a proclaimed ideal, to representing things the way they might be or the way they ought to be (e.g. Edmé Bouchardon's equestrian statue of Louis XV as a Caesar). Finally, artistic reality regards the representation of things the way they are (e.g. Jean-Baptiste Pigalle's sculpture of the aging Voltaire with all the deep wrinkles of his nude body). Whether we are the observer or the maker of art, the layered relation between art, nature, artistic truth and artistic reality is essential not only for an occasional interest in meaning, but more importantly, for our living within the world of meaning. Constructing the world meaningfully requires its symbolization.

Symbolic thought brought some important consequences for the theory of knowledge. In order to integrate the world into our thinking we require a mediation: to make an object we require that one object stands for another. This implies resemblance, an imitative resemblance in which art expressed and represented the characters of greater gods and lesser gods, the characters of kings or peasants, the characters of the church's cardinals and the character of virtues called cardinal. Thus the arts could personify continents, nations, and their liberal arts and sciences. They could fully express the complex range of human emotions, joy or sadness, pity or ruthlessness, nobility or shame, grace or clumsiness. When a visual art represents it denotes a relation between an idea and an image, or a relation between an image and another image by virtue of which the new image is said to be "of", or to "stand for" that idea or that other image. Representation is adequately known when an individual and a society understand what it represents, provided that the artist skillfully executed the work to make the subject of the representation clearly evident.[1] In other words, representation may be said to be successful when the effects intended by the artist (intrinsic properties such as artistic composition) correspond to the effects on the perception of the observer (the level of understanding). Artistic representation requires a deliberate resemblance between a crafted image and a subject.[2]

Expression can be said to *emanate from* works of art. It gives to society forms and meanings drawn from within art itself (larger artistic conventions) and from within the artist's mind (self-expression). If a painting expresses, it means that the artist willed this expression and the observer recognized it. Representation can be said to operate in the opposite direction than expression. It *draws from* society forms and meanings with the full expectation that works of art will embody or "stand for" these forms and meanings. If an artwork represents, it means that the artist willed a composition based on meanings drawn from society and that the observer recognized them. Naturally, artistic expression, representation, and social meanings evolve through reciprocal influence. Expression and representation need not be categorically separated even if they imply different directionalities. Expression and representation operate dialectically in the mind of the maker and that of the observer. Together, they affect additive transformations of the artistically given drawing meaning from society and giving meaning back to society taking art as a departing point.

On the level of artistic production, of expressive qualities, imitation and invention—that inseparable couple which made for art's rich history—produce the resemblance of an object within another object that becomes its image. Together artistic imitation and invention transform pre-existing forms, adapting them to different contexts; and whereas invention brings something new, every novelty is not necessarily inventive. On the level of meaning, imitation and invention mobilize an ensemble of symbolic relationships that allow the content of one object to reveal another. Where there is artistic making (*poiesis*), there are the two entwined concepts: imitation and invention. Where there is symbolism there is imitative transparency.

---

1 Theories of representation have explicitly dealt with intention, content, and perception in art. Among them are the illusion theory, the resemblance theory, the seeing-in theory, and the semiotic theory. See R. Wollheim, *Art and its Objects*, (Harper and Row, 1968); N. Goodman, *Languages of Art*, (1969), (Hackett, Indianapolis, 1976); E.H. Gombrich, J. Hochberg and M. Black, Eds, *Art, Perception and Reality*, (Johns Hopkins University Press, Baltimore, 1972); A.C. Danto *The Transfiguration of the Commonplace*, (Harvard University Press, Cambridge, 1981); M. Baxandall, *Patterns of Intention: On the Historical Explanation of Pictures*, (Yale, New Haven, 1985); K.L. Walton, *Mimesis as Make-believe: On the Foundation of the Representational Arts*, (Harvard University Press, Cambridge, 1990); P. Alperson, *The Philosophy of the Visual Arts*, (Oxford University Press, 1992); N. Carroll, *Philosophy of Art: A contemporary introduction*, (Routledge, 1999).

2 In this sense Aristotle's observation that there is no thinking without the image fully obtains. Reason bases its activity on the images furnished by *phantasia*, and these images bridge the distance between sensations and concepts. For Aristotle's thinking through images and phantasia see his *On the Soul* and *On Memory and Reminiscence*. Some modern translators and commentators of Aristotle's work have chosen, unfortunately, to discontinue their translation of *phantasia* as imagination, preferring instead the word representation.

The technological order

This condition obtained until the advent of artistic modernism and the pervasive phenomenon that Jacques Ellul termed *la technique*, upon which all cultural productions, including art, architecture and the city, and much of the natural environment have been lavishly wagered. We shall return to Ellul's definition of *technique* in a few paragraphs. To continue our discussion of artistic making, prior to modernism imitation meant that artistic objects are made out of combinations of other artistic objects, cities and buildings are made out of combinations of other cities and buildings, while invention sought to improve the rational choice made from exemplary precedents. Whereas skepticism regarding the practice of imitation as part of a historical continuity began to be voiced in the eighteenth century, it is important to note that imitation and invention, in general, were considered as two facets of the same coin well into the nineteenth century and increasingly again since the 1980s on the part of modern traditional artists and architects. With modernism, however, invention became an end in itself. The different facets of the same coin: imitation and invention now became two identical facets: invention and invention. This separation was given currency and legitimacy by modernist critics and art historians who wrote histories of art as histories of artistic ruptures. The sequential passage from Mediaeval to Renaissance art, to Baroque, to Neo-classical art, to Eclecticism, to Modernism, was assured by rupture, and artistic invention was identified as the very cause of this rupture. Thus, the coupling of rupture with invention came at the expense of uncoupling imitation from invention. Quite significantly, rupture and invention in the arts and architecture came to be conflated with the idea of progress in the sciences and technology. No distinction was made between artistic invention and techno-scientific invention.

Artistic and architectural production was now considered to be *all invention* at the same time that imitation and invention came to be understood as antagonistic rather than complementary concepts. To be inventive meant that artists and architects were to practice *creatio ex nihilo*, the making of objects out of nothing, following their individualistic expressionism—a *tabula rasa* that excludes artistic precedence or tradition with a sleight of hand. Modernist painters, sculptors, and architects ostensibly gazed on an empty canvas and materialized forms that have not been seen before. They did so, we are told, under the impetus of a supra-personal technological *zeitgeist* that they were singularly gifted in incarnating. The artistically modern not only succeeded previous art forms, it violently confronted them and rejected them. The relation between artistic making and the very context in which artistic making occurs were meant to be in a permanent state of crisis; and this state, insisted the protagonists of modernism, is the exact reflection of modernity. Opponents of modernism insisted that there was a difference between the crises within artistic modernism and the fact that artistic modernism itself constituted a crisis of art. Society itself had to be reshaped in order to reflect the new technological order, an order that will liberate humanity from the socio-political shackles of the past as well as the art and architecture that represented them. Because of faulty associations between condemnable political conditions and the art and architecture that were produced during these conditions, entire artistic traditions were thereby condemned. Furthermore, modernism considered the technological age to be the ultimate paradigm that gave meaning to humanity's visual and verbal cultures, and modernist art and architecture were the uniquely appropriate symbolic forms for this age.

If imitative transparency characterized visual and verbal production for millennia, modernism replaced this transparency with a dense opacity to external meaning as well as to previous artistic conventions.³ Only, artists and architects do not create in the elementary sense of creation from nothing as their forms are invariably based on older forms even if they are the inversions or abstractions of previous forms. Instead modernist forms have been made, situated, evaluated, and judged with respect to *technique* as the value of all values. The big contradiction resided in the modernist claims to freeing the imagination and invention while wholeheartedly accepting technological determinism. Yet, despite their fervent wish to be unique and produce the previously unseen, and despite their determination to separate imitation from invention, modernist artists and architects still learned, appropriated, and practiced their preferred forms through undeniable imitative acts for two important reasons. First, any collective construction of artistic or architectural qualities and forms and their transmission over several generations means that a tradition is being elaborated. Second, artistic and personal identities are inextricably connected to those of other architects who share the same world-view. For these reasons modernism itself became a tradition. At one point, even a renewed avant-gardist urge toward continual change passes from being a transitory phenomenon to becoming an established practice, even if only for the duration of a few decades. Those who denied tradition themselves developed into a tradition.

Opponents of artistic modernism assailed its fundamental bases in historicism (in historical determinism), in the cult of the *zeitgeist*, in the failure to distinguish between modernity (in a temporal sense) and modernism (as an artistic ideology), in the symbolic poverty of modernist art, in industrialized mass production, in artistic abstraction and its remoteness, in the profound alienation felt in urban contexts where modernism dominates, and in the insistence on glorifying technology (especially amongst architects) no matter the consequences. Here, Ellul's contribution points to two important conclusions: 1) that far from being the simple expression of artists who use technology at the service of their art, modernist art was instrumental in inducting the mind into what he called "the technical system;" and 2) that this phenomenon caused the devaluation of the image and the word—the two expressions, *par excellence*, through which social life is invested with meaning.

Ellul, technique, and art

The fact that Ellul published two studies dedicated to the image and the word—*L'empire du non-sens* (*The Empire of Non-Sense*, 1980) and *La parole humiliée* (*The Humiliation of the Word*, 1981)—almost simultaneously is evidence of his deep concerns about the devaluation of meaning that befell contemporary visual and verbal cultures. Emptied of meaning, of symbolic function, the image and the word had both been humiliated inside a technological milieu whose far-reaching intervention disrupted humanity's pluralistic ways of artistic making. It is noteworthy that Ellul's *The Empire of Non-Sense* and *The Humiliation of the Word* were released following the publication of two of his seminal works on the phenomenon of *technique*: *La technique ou l'enjeu du siècle* (*The Technological Society*, 1954) and *Le système technicien* (*The Technological System*, 1977). His third study, *Le bluff technologique* (*The Technological Bluff*, 1988), completed his life-long reflection on one of the most monistic forces ever constructed: the framing of the mind in a technological direction and the restructuring of multiple visual and verbal products in order to fit that direction. *Technique*, in Ellul's thought, is to be distinguished from technology. *Technique* designated the pursuit of utmost rationality and efficiency

---

3 On Conrad Fiedler's notion of opacity in works of art see Philippe Junod, *Transparence et opacité* (1976), (Jacqueline Chambon, Paris, 2004).

that can be attained at any given moment and the consequent conquest of all areas of nature and human endeavor for this purpose. Technology designated the many possible discourses on *technique*, e.g. sociological or philosophical studies on *technique*.[4] Thus, in Ellul's sense, modernist theories of art and architecture are an apologetic discourse on *technique*. Modernist theory was a techno-logy.

In a poignant analogy Ellul once remarked that if one were travelling on a train then one could not see the direction that the train is taking. One must disembark from the train of *technique* in order to gain a perspective on its direction and effect decisions from outside its empire. Such a task is truly formidable considering that *technique* as a system (*le système technicien*) plays a determining role inside society, a role that participates in steering the principal forces of this society toward a technological direction, a direction that always appears inevitable to the technologically-formed mind.[5] One of the salient characteristics of Ellul's *Empire of Non-Sense* is that his critique of modernist art was based more on the texts that justified modernism and less on modernist art itself. He is less concerned with the clusters of forms and positions elaborated by several artistic and architectural movements that include Constructivism, Futurism, Cubism, De Stijl, Expressionism, the Bauhaus, Functionalism, the International Style, or the declarations of C.I.A.M. congresses, and more with the fact that they were all informed by *technique*, and that they in turn validated the technological milieu. Ellul goes to the heart of the matter: modernist art theory is not only a justification of modernist art forms; more importantly, it serves as a justification of the technological society. In keeping to his train analogy, he engages modernist art from the 'outside'. And while he also offers a genuine critique of modernist art, he is unwavering in his judgment that modernist art and its theory are justifications for the integration of "humankind into the technical complex." This characteristic sets him apart from others who opposed modernism from the 'inside', that is, on the grounds of art theory and architectural theory.

Artists, architects, and their critics, apprehend and make the world imagistically, and they apprehend and make modernity imagistically. Put differently, their understanding of the world is strongly mediated by the images that inhabit the world and the images that inhabit their minds. Ellul, by contrast, is a man of the word whose sensibilities are more inclined toward symbolic content, to the meaning that should underlie artistic form and justify it. Much of his understanding of the world is mediated by the word, and less so by the image. In fact Ellul was quite alarmed by the invasive proliferation of images in the technological society. His strong Protestant aesthetics played a significant role in this distress, which he expressed as a religious conflict between the image and the word.[6] But Ellul is not an indiscriminate enemy of visual culture. He was most concerned about a particular kind of image, a triumphalist image whose empire humiliated the word, namely: the "technicist" image that

---

[4] It is important to note that Ellul's critique entails neither passing indiscriminate judgments on anything technological, nor a simplistic techno-phobic confrontation between humans and machines. While the machine is the most obvious manifestation of the domain of *technique*, it is only one among its many phenomena. *Technique*, for example, is the mentality that divides all inhabited territory, e.g. agriculture and the city, into a collection of separate mono-cultures and mono-functional urban zones. Mono-cultures and mono-functional urban zones relate to the rational and efficient abstractions of a technological system that radically transforms natural cycles and the nature of the city. To illustrate this, consider the eradication of a previously pluralistic biodiversity by agricultural areas dedicated to one single crop. Consider also how the city comes to be divided into zones dedicated to the exclusive uses of commerce, or culture, leisure, industry, and to dormitory suburbs, thus rendering dysfunctional the city's previous plural coexistence of all the functions of life within an optimal area based on human measure. Thus zoning regulations, in Ellul's understanding, are a discourse on *technique* as applied to natural and urban territory.

[5] For Ellul's discussion of the technological system as an autonomous and totalizing system qualified by an absence of finality see his *Le système technicien*, Calmann-Lévy, 1977.

[6] See his *La parole humiliée*, Seuil, 1981, pp. 202-224.

frames the minds of citizens. Citizens of the technological society were consumers of technicist images—images that were justified by an ideology that glorified presentness as the leading edge of modernity. "With the ideology of instantaneity in art, with immediacy, with spontaneous creativity (the happening, etc.), we are in the presence of a pure assimilation into the technological processes, and a total negation of all that has been considered art since the beginning."[7]

Artists and architects, we said, apprehended the world with images and made the world with images. This, however, is not to say that artists and architects are not concerned with meaning or with symbolism. Indeed some are acutely concerned with meaning. Only, as makers of visual culture they place a higher value on the image, the visual form. Yet, in a predominantly modernist culture, the overriding purpose for which artists and architects produce forms has more to do with self-expression than a contribution to the public realm, the sense-in-common, or the general good. This phenomenon takes particular importance with respect to the idea of meaning in art and architecture because modernism inherited and amplified the Romantic belief in the artist or architect as a solitary genius who walks in no one's shadow and who produces forms that have not been seen before. The modernist rupture and transgression, in Ellul's terms, of previous traditions assured a *tabula rasa* while at the same time exponentially exalting the personae of artists and architects by putting at their disposal all the massive means of technology. Iconoclasm and shock-value worked hand in hand with the belief that the artistic value of modernist art is meant to increase while simultaneously debasing that of traditional art. One artistic form is exalted by humiliating another. The theoretical justification of modernism shifted the artistic intent of elaborating a tradition—ever a collective endeavor—toward a deepening interest in the artist's personal life which itself came to be considered an object of art. Here we have a replacement of art by the artist, as the artist became a sacralized figure, a priest in a secular religion whose genius must always be valued and whose decisions are almost beyond judgment. Even the empty canvas became an object of art—itself a mute comment on a painting that could have been realized. And yet, the act of withholding a painting from manifesting came to be endowed with the aura of art, as if this intensely private act was precisely the reason why it should matter for culture at large—a condition of no sense. Here, the newness of this gesture, that highly coveted quality, was achieved by the very absence of artistic gesture. In exasperation Ellul protested that "to apply exactly the mentality of Epicurus is no aesthetic creation."[8]

With positions such as these, the frenetic pursuit to distinguish oneself, especially when undertaken by a considerable number of artists and architects over several decades, amounted to an exclusion of the artistic sense-in-common in favor of the self-referential sign. The artistic sense-in-common here is distinguished from common sense because common sense could be applied by simple habit. By contrast, sense-in-common designates sets of artistic conventions whose justification derives from the continual reflection, agreement and disagreement between many free minds contemplating the same artistic concerns, and enriched by the wisdom of experience. This condition has been violently reversed in modernism, particularly among architects who frequently put self-expression over and above the idea that architecture as a public art is called to serve the City, the public realm, the *res publica*.

---

7 *Ibid*, pp. 249-250. My translation.

8 *L'empire du non-sens*, p.34. My translation.

Modernism and progress

Ellul was appalled by the complicity between modernist artists and the entwined milieu of business and politics, the museal institutions and their critics, as well as the mass media that served them. Their conjugated effects made for a cultural atmosphere where the admiration of modernist art was obligatory. He was little affected by the propagandistic sophistries of modernist art theory because he saw modernist art forms as technological forms situated within and explained by a society that is meant to be technologically determined in the first place. Modernist art and architecture and their theory sought to form and conform the mind according to a technological *weltanschauung*. Previous forms and traditions that have been painstakingly elaborated and layered over centuries within a cultural sense-in-common could therefore be iconoclastically discarded. Modernism had become a monistic force that was justified by art and architectural historians and critics as if it were a historical necessity, a panacea toward which all previous artistic production was unalterably led and from which it definitely separated. Classicism's old belief in an unsurpassable past artistic ideal was replaced with the belief in a future ideal that will somehow arise from a historical contingency determined by *technique*. Apologists of modernism ardently argued for this belief, and some of them, like several Futurists, argued with shocking violence. In so doing, they produced conflations with far-reaching consequences, among which is the 'conflation' of teleology with progress, as various historians of art and architecture wrote this onflation into their narratives.[9]

Progress differs from teleology in the sense that teleology does not necessarily imply improvement. A *telos* (Greek: goal, end) might very well lead a chain of events toward undesirable conclusions. Such, for instance, is the difference between promise and progress. In their good aspirations early modernists in art and architecture sought to wed their preferred artistic and architectural forms to progressive social ideals and their beliefs in the redemptive role of technology with the full expectation that historical events will gradually unfold in the direction of their goals. Yet, the decades that followed showed that modernist art and architecture became tools of daily market forces having little to do with earlier stated ideals, while the unrestrained belief in technology led to catastrophic environmental consequences and a long-standing unwillingness to admit these consequences. Progress is a particular way to represent historical time that differs from the simple notion of development in that progress advances toward a certain finality, a better finality. Progress implies that history moves according to a unified direction, and that historical periods constitute the various steps of that progress in which a principle gradually realizes itself and justifies all the changes. For Jacques-Bénigne Bossuet, this principle is God governing history; for Voltaire and Condorcet it is Reason accompanying history; whereas for Hegel, Reason systematically justifies the progressive movement of historical periods on their way to the realization of the Concept. Historical events or periods gain their significance depending on the place they occupy

---

[9] For example, the work of historians: Emil Kaufmann, *Von Ledoux bis Le Corbusier*, (1934), (French translation 1994). Sigfried Giedion, *Mechanization Takes Command*, (Oxford University Press, 1948); *The Eternal Present: a contribution on constancy and change*, (1962), (Princeton University Press, 1981); Nikolaus Pevsner, *An Outline of European Architecture*, (1948), (Penguin Books, 1968); *Pioneers of Modern Design: from William Morris to Walter Gropius*, (1949), (Yale University Press, 2005); *The Sources of Modern Architecture and Design*, (Oxford University Press, 1968); Henry-Russell Hitchcock *Architecture: Nineteenth and Twentieth Centuries*, (Penguin, 1958); Leonardo Benevolo, *The Origins of Modern Town Planning*, (Routledge & K. Paul, 1967); *History of Modern Architecture*, (Routledge & K. Paul 1971); *The History of the City*, (MIT Press, 1980); Manfredo Tafuri and Francesco Dal Co, *Modern Architecture*, (1976), (Harry Abrams, N.Y., 1979); Kenneth Frampton, *Modern Architecture: a critical history*, (1980), (Thames & Hudson, 2007).

within a unified and progressive chronological development. Consequently, progress implies the merging of meaning with direction.

Progress has now become such a routine belief that it passes unreflectively for a historical given. Progress for artists, and especially architects, has been deeply entangled in means; and when the technological means proliferated, Ellul reminds, the ends for which the means were developed disappeared from sight. But the post-modernist self-conscious reaction against the modernist justification of progress was not embraced in all cultural spheres. When some thinkers saw the weakening of the Enlightenment certainty regarding the progressive direction of history, they concluded that this was the dissolution of history itself.[10] Others went further, arguing that the acceleration of events has proceeded so exponentially that it is now beyond our capacity to see them as history. Others still went as far as to propose that the immense network of products and signs within the consumer society makes it such that we can no longer distinguish historical reality from the myriad consumer images that occupy the reality of experience.[11] The multitude of images that now inhabit the technological consumer society have the power to condition contemporary understanding to such a point that they already frame the intellectual assessment within this society becoming a kind of lens through which intellectuals look both at past and present cultural forms. Accordingly, the mind is strongly affected if its grasp of the present-as-history is enclosed within this context. Paradoxically, although modernists championed their work as a decisive rupture from historical precedents, they nonetheless cherished the idea that they were carried by inexorable historical forces to the point they presently wish to occupy. For reasons such as these, many artists and architects rebelled after decades of proscriptive modernist control on artistic forms, on their history and their explanation. One of the first rebellions, since the late 1970s, rose to oppose modernist determinism by calling for a cultural milieu that accepted plural artistic expressions, a milieu that was characterized by its openness to the lessons of previous artistic traditions, a milieu that is generally known by the cumbersome term, post-modernism.

The retreat of symbolic thought

The idea of technologically remaking the world, the complex set of phenomena that Ellul called *la technique*, had crucial effects on architecture for the simple reason that architecture constitutes one of the two environments in which we dwell: the city and nature. Transcending mere building and the mere sheltering of human habitation, architecture serves the purposes of the public realm (library, municipality, court house, church, museum) and the private realm (house, offices) by endowing them with their proper expressions. But, remaking the world technologically was conflated by modernist architects with the uncertain belief in architecture as a scientific discipline. This idea operated on the assumption that science (understood as technology), architecture and art, were linked by the *same* idea of progress. Whether it is cities, buildings, ocean liners, automobiles, aircraft, furniture, or kitchen utensils, the technological society was to be made with technological products and be represented by these same products. Every product must be qualified by a technological character. This unassailable belief exerted some far-reaching influences on symbolic thought, on artistic expression, on architectural character, and on the art-language and architecture-language analogy.

10 See Gianni Vattimo, *La fine della modernità*, Garzanti, Milano, 1985.
11 See Jean Baudrillard, *Simulacres et simulations*, Galilée, 1981.

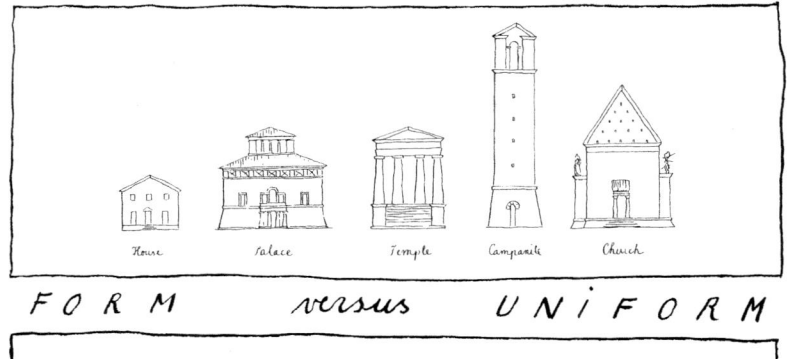
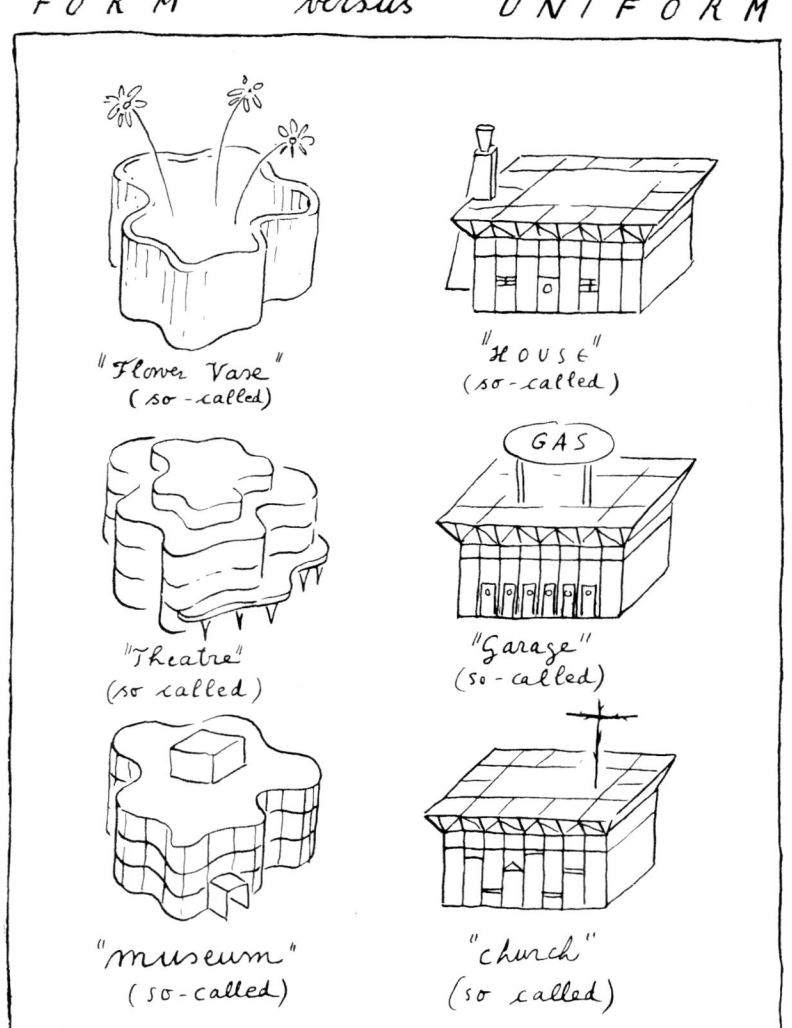

Fig. 2: Léon Krier, *Nameable Objects*

Because technology was both the symbol and the product, the true and the real, the signifier and the signified, the artistic idea and its representation converged or rather collapsed into each other. Because meaning was internal to *technique*, it became enclosed within a self-organizing and self-referential system that accepts no external feedback. It became non-dialectical, a presentational immanence—a "spurious infinity" affirms David Lovekin in his use of the Hegelian expression.[12]
In the technological system that permeates society, the idea of making always resembles itself and replicates itself. It became its own ends. For this reason the technological mind became monistic. It also dismissed the symbolic ends, forms, meanings, and cultural conventions that previously allowed art or architecture to express a civic character or a private one. And yet, although modernist architects enthusiastically embraced the non-dialectical modes of the technological system, they still wished their forms to symbolically represent the technological order because they still retained the traditional idea that any object acquires a symbolic function simply because it was made. They justified their architecture as a *reference* to technology, while in reality it *was* technology. So the problem was not that there was a lack of correspondence between "image and substance"[13] but rather that the image and content were equal. Thus, what is usually considered to be one of modernism's strongest points, that is, the view that art and architecture symbolized the technological society and its informing *zeitgeist*, is actually its weakest. A symbol that recoils onto itself is a vicious circularity. A symbol that symbolizes itself is a condition of no sense.

> *If artistic making in the poetic order is qualified by a purpose other than itself, the artistic making in the technological order has itself as its own purpose. A significant artistic stance then resides in not only distinguishing, but also in choosing between the poietic order and its art and the technological order and its art.*

The symbolic function received another setback with modernism's attempts to eliminate the difference between the imitation and the copy while producing numerous identical repetitions of technological buildings and products in every continent irrespective of the character of place. The exorbitantly anti-ecological steel and glass skyscrapers that dot the planet as one of the sacred images of modernist progress bear little belonging to any place. They are built in every continent while belonging nowhere, yet always appearing necessary as one of the sacred images of modernity.

Eliminating the difference between imitation and the copy also meant eradicating the distinction between the type and the model. Architectural types collapsed into technological standards, e.g. the skeletal structure of the *maison domino* was meant to be the standard underlying the very idea of every modern building. Because any architectural character can be attached to this skeletal structure, structural form can be dissociated from architectural character and meaning, which in turn become removable attributes. In such a way artistic truth is displaced. If any architectural character can be attached to a mute skeletal structure then the result is kitsch—one of the most abundant phenomena of the technological society as Léon Krier has tirelessly repeated for several decades.[14] This phenomenon

---

12 See Lovekin's *Technique, Discourse, and Consciousness*, Lehigh University Press, 1991, pp.98-105.
13 As suggested by Robert Venturi, Denise Scott Brown, Steven Izenour, *Learning from Las Vegas*, (MIT Press, 1972), p. 137.
14 See Léon Krier, *The Architecture of Community*, Island Press, 2010.

is most evident in the confusion of genres that abounds in the technological society where a warehouse with a cross on its roof conveys that it is a church, where an amorphous and sinusoidal vase might also be the shape of a theatre, a library, or a museum. Thus, when ordinary citizens engage in caricatural naming of buildings, architects ought to listen because naming calls forth an object's nature, its character. Naming lays bare an object's artistic truth. Thus, designating the Centre Pompidou in Beaubourg in Paris as an "oil refinery," or the new museum for the Ara Pacis in Rome as a "petrol station" shows that the general public possesses an indelible sense of what architectural character "ought" to be even if this public may not necessarily know the exact form this character may take. When artistic shapes and architectural shapes are exchanged and dissolved inside a technologically determined reality a crisis of meaning is precipitated—a condition of no sense.

Space and visuality in modernist art, architecture, and also music, were expressions of technological operations. In many ways the empire of technological means exploded the limits or boundaries between the arts. Architecture could become sculpture and vice versa, while architects transformed cubist paintings into the plans, sections, and elevations of buildings following the example of modernist prophets such as Le Corbusier. The keyboard of an electric organ produces the sound of drums and cymbals. An artist who produces 'art work' through a collage of unrelated photocopied images with varied colors is evaluated on the same level as the painter who composes and proportions a painting with the painstakingly judicious use of the brush following years of assiduous training and introspection. To a technicist mind, the photocopier and the brush are both means that are equally received irrespective of artistic skill; and the technicist mind considers the proliferation of means to be a necessary condition of artistic freedom. Only, this proliferation of means, Ellul insists, makes for a freedom from which the artist cannot escape. In this triumph of means any combination of forms becomes possible. Artistic genres, or traditional modes of artistic composition, are considered obstacles in comparison to the emancipatory and seductive technological means. Yet, contrary to prevalent belief, technological means do not necessarily facilitate the expansion of artistic freedom, nor the quality of art. If the manifestation of artistic form previously depended on a symbolic thought that instantiated expression and representation through manual skill, this manifestation has now been replaced by technical processes and operations and the near elimination of what has hitherto been known as symbolism, whether it is art imitating nature, or symbolizing religious themes, or social mores. It is important to note that the *augmentation* of technical means has been accompanied with a *diminution* in symbolic form and meaning. Thus the distinction between an object of art wrought with skill and the multiplication of technological processes and products has been blurred. Here we encounter one of the greatest paradoxes of the technological society: on the one hand, the proliferation of objects imply the triumph of the object, on the other, this very proliferation also means the obsolescence of the object—a condition of no sense.

L'empire du non-sens can be considered *un cri de peur* on the part of a man who laid bare his fears and disquieted concerns about a society so utterly permeated by *technique* and so docilely accepting of this invasion. Artistic creativity, or invention, were "radically and totally integrated into the technical system,"[15] and this integration passes almost unnoticed because the technological society drowns

---

15 *Empire*, pp.30.

within a pandemonium of noisy objects and images that have been emptied of meaning. The sheer proliferation of technological objects and images makes it difficult to imagine another world, while modernist art and architecture affirm and confirm the technological society on a daily basis. The result is a disenchanted world because many of its symbolic functions have been suppressed.

L'empire du non-sens was published in 1980, and although opposition to modernism in art and architecture was beginning to be expressed in the 1970s, Ellul could not account for the solid alternatives to modernism that developed since then. Even if the teaching and the practice of art and architecture today remain predominantly influenced by modernistic forms (the technicist image) there are glimmers of hope that one discerns in artistic academies and in professions. Several art schools and ateliers around the world (e.g. The Florence Academy of Art, and the Angel Academy of Art, also in Florence) have now emerged where the study of nature, the human figure, beauty and proportions, landscape painting, historical subjects, realism, form the core of their curriculum. A handful of architectural schools and private institutions dedicated to traditional architecture (e.g. the University of Notre Dame, The Prince of Wales's Foundation, The International Network for Traditional Building and Architecture, the Institute for Classical Architecture) are now established. They teach traditional architecture and urbanism in view of constructing an enduring world where nature is seen as the enclosure, where the city is built inside of nature, and where architecture is built inside the city, in that hierarchical order. Parallelling these academic developments, many painters, sculptors, architects, musicians, poets, are now practicing the humane art of dwelling wisely on this planet based on the successful lessons of past experience and on the avoidance of past disasters. Following the hard-earned lessons since the Enlightenment, the practice of tradition will benefit by avoiding a blind faith in an unsurpassable and idealized past, and a blind faith in an unknown idealized future that will somehow emerge from a technologically determined reality. As Ellul himself acknowledged, there is much in human nature that refuses to be integrated into a technological system that frames the true, the factual, and the possible.

ART AND TECHNOLOGY
All or Nothing

David Lovekin

Jacques Ellul, critic of technological dominance, announced in his magnum opus *The Technological Society*, that art and morality went the way of the ruffled sunshade on McCormick's first reaper. When concerns impeded a rigorous and rational productivity, they were sidestepped, transformed, or simply abandoned. This is a compelling metaphor, and like all metaphors, hides as much as it reveals. It, first, suggests that art is either a disposable decoration or a supplement and/or support for technology, which is here pictured in machine usage. To many current readers, this claim would seem absolutely false. Art appears to be in no danger and is present on any street corner—billboards, graffiti, weekend art and craft displays; it is present in every magazine, on every television program, on every radio station. Art is taught in public and private schools, collected in galleries and museums, and with the ubiquitous smart phone, everyone becomes a photographer-artist. We have no end of images before us. A majority of American youth—adults too—seem to be in training for stage and screen—at least in preparation for You Tubing and Karaoke Bar outings. Reality television programs suggest that television has become reality.

These observations, which appear negative, perhaps hide an elitist agenda. Do these examples prove that art is proliferating or that it is being diminished? We can easily imagine the next question: who's to say what is or what is not art? How can this be answered without question begging? If no one can say, then silence should reign on all accounts. If only artists can say, then who are the artists? Ellul would say that, currently, no one seems to be able to say, short of museum curators, critics, and other technicians trained in art history and marketing procedure, and, it is here that Ellul locates a problem. In pre-technological cultures (allowing Ellul's special understanding of that phrase) many could say: communities could say; those trained in the arts could say; those seeking to honor the gods and traditions could say. Like Louis Armstrong said about jazz, "You know it when you hear it; if you don't you won't." (This is, of course, a paraphrase, or an improvisation, but liberties are sometimes required.)

I

Jacques Ellul (1912-1994) published at least fifty volumes and over 1000 articles in newspapers and assorted journals, but, nonetheless, regarded himself as an author of one long book with a variety of chapters.[1] His writings included studies in the history of law and social institutions (he was a professor in the law faculty at the University of Bordeaux from 1944-1994), studies in theology and biblical criticism (he was a lifelong member of the Reformed Church of France but remained a frustrated minority voice in theological and ecclesiastical matters), and studies in social and ethical criticism (*The Technological Society* was published in 1954 and then translated into English in 1964 with Aldous Huxley's strong recommendation and featured a glowing foreword by distinguished American sociologist, Robert K. Merton). Such diversity gives pause.

Ellul outlined defining life moments that focused his analyses in relation to an encounter with a being so great, with a Wholly Other, while translating *Faust* in his mid-teens,[2] to a reading of Marx that explained the economic-political situation of the 30s and that continued to play out in the reification of commodities in modern life,[3] and to an understanding—accomplished with his life-long friend Bernard Charbonneau—that technology was the major force confronting the modern world. All institutions and traditions were eclipsed by this force, which he came to understand as an intention, a world-view, a way of being in the world, a mentality. Mentalities are situated, as was Ellul's. His notion of technology went well beyond mechanization and the use of tools, methods and means, and was an Ariadne's thread to all of his writings; technology is expressed in the whole of culture and, most specifically, in communications and in the political dimension.

Jacques Ellul's *The Empire of Non-Sense: Art in the Technological Society* was published in 1980 by the *Presses Universitaires de France* as part of Lucien Sfez's series *La politique éclatée* (*Policy Exploded*). The book jacket announced that this series brought authors together with the common purpose of "… not dividing or closing off analyses in what would conventionally be called policy ("*La" politique*) but which would pursue the other areas of its scattered fragments." Ellul's study is about art, but more specifically the art of a technological society, for which there was no common measure. From Ellul's perspective, politics had become an illusion that required propaganda and art to construct a *sensus communis*, a sense in common no longer expressed by cultural traditions. Art became the diffusion of policy, *la technique*. Technical consciousness lost a symbolic sense required in a reach for value beyond fact, meaning beyond measure, and an aesthetic sense that in the past formed a basis for making, for *techné*, understood as know-how.

In *L'Illusion politique* (1965) Ellul warned that *le politique* had become *la politique*, that *le politique*—a concern for goals and values that portend the common good—had devolved into *la politique*—a concern for procedure and method and implementation.[4] The illusion of politics was made possible by the cliché that politics was all-important when it no longer existed. True politics (*le politique*), Ellul believed, had collapsed into the show of politics, where image became reality. He stated:

> *The only domain in which politics can still act is in the domain of current events, i.e. the sphere of the ephemeral and the fluctuating. As a result, the feeling that a political decision is truly serious has been lost. What becomes visible is no longer anything but appearance.*[5]

Ellul was thinking of the newspaper and the mass media as the purveyors of the visible and the ephemeral. *La politique* was only possible when reality was reduced to the here and now, to an image, the visible, to that before which we stand. He considered that the modern state, tied to technological development, eclipsed tradition. Ellul announced that Hitler had won the war, that all modern states had come to apply his methods of unifying technological development and of using propaganda spread by the mass media.[6]

Ellul believed that the dimension of *le politique* was not present in the here and now but was what that present sought—the Beautiful, the Good, and the True. From the present current-mindedness of what Ellul called *la technique*, these values were the ephemera of tradition that hindered the progress of modernity with a new sense of "necessity," which, Ellul claimed, became the current ephemeral.

If we are placed in a period of history and in a society in which necessity becomes ever more exacting, then nothing is truly continuous or durable. Our entire civilization is ephemeral. When one glorifies increased consumption, one must discard machine-made objects in the course of rapid usage. We no longer repair things: we throw them away. Plastic, nylons, are made to be used for an infinitesimal period of time and, as they cost nothing, are destroyed as soon as the gloss of newness is gone. Houses are constructed for the duration of their mortgage; automobiles must be replaced every year. And in the world of art we no longer build cathedrals, but we make moving pictures, which—though real works of art in which man has fully committed himself and expressed his most profound message—are forgotten after a few weeks and disappear into movie libraries where only a few connoisseurs can find them again.[7]

Traditional necessity became a necessity denied, but not without problems. Materials were made that lasted "for an infinitesimal period of time," the embodiments of absolute efficiency and rational organization—what will define *la technique*—and yet were to be discarded. Values were needed and some absolutes were required—the new ephemeral must be good, valuable, and appealing, but not for long. Time and memory had to be transformed.
Ellul wrote:

The man who lives in the news … is a man without memory. Experimentally this can be verified a thousand times over. The news that excited his passion and agitated the deepest corners of his soul simply disappears. He is ready for some new agitation …. This means that the man living in the news no longer has freedom, no longer has the capacity of foresight, no longer has any reference to truth. Lequier has said that "memory is free man's action when he turns to his past acts in order to retrieve them." Memory in a personality is the function that attests to the capacity of acting voluntarily and creatively; personality is built on memory, and conversely memory lends authority to personality. "One only remembers oneself, which means that one must already be self, capable of creating oneself in order to remember."[8]

Memory and freedom are linked to the individual's grasp of what he or she is and is not—a being in time and place in which past, present, and future provide constituting senses of awareness and meaning that enable a choice that expresses freedom. A present with no past or future is no present but only a place in the empire of non-sense.

We find in Ellul's writing continued references to art and to the importance of a sensibility able to transcend technique in the political domain. In his own life we can recall the importance of *Faust* and of a Marx who attempted a challenge to bourgeois sensibility and of a sense of the Wholly Other. We should note that his mother was a painter and that he produced two volumes of poetry toward the end of his life. On the last pages of *L'empire* he wrote:

By necessity, the politician manipulates a double discourse: one discourse is objective and meant for the inner circle of party members to reveal what he will truly do at the level of pure praxis and a Machiavellian strategy to attain ends, and the other discourse, directed toward the public, is clothed in absolutes aiming for a groundless reality promulgated by Art and the mass media, of course! Only a radical challenge to the political system will enable the reestablishment of meaning and value beyond those tricked out by propaganda. But, this challenge to politics does not imply a retreat to the "ivory tower": to the contrary, symbolizing the reality of today puts in motion a force that questions the reality of tomorrow, which would lay the groundwork for another type of society. This was the creative activity in art each time it fulfilled its plenary function that is the fundamental recognition of what is (the Janus face of State and Technique) in the name of a new birth. Art is the procreator, first, and then the midwife. But, to that end, it is necessary to shake off the general defeatism of all modern artistic activity for the past century.Art would now assume the task of encouraging man to rise up against the servitude and progress tied to consumerism, to stir up the ever-present burning desire for the absent Wholly Other.[9]

The Wholly Other was present in its absence for Ellul. He found it in what Technique was and was not; he found it in a Christ who was beyond institution and dogma; he found it in the power of the symbol—the word—which was the attempt to grasp the Beautiful, the True, and the Good against whatever odds prevailed but which had been humiliated. Ellul's examination of *la technique* was an attempt to put into words ideas that went beyond facts but that explained facts to form and to reach a new collective memory. Ellul's belief that the technological society denuded memory and meaning; seeing technology was not a neutral force of using tools and means to accomplish ends that would bring about progress and prosperity and unlimited freedom but was a form of amnesia that founded and furthered technology itself. Within the technological mentality, art was no longer possible.

II

Art in traditional societies made symbolic gestures toward the Beautiful, the True, and the Good, goals the modern artist considers passé. The modern artist is deemed free, a veritable revolutionary, beyond tradition. The modern artist might embrace the ugly, the false, and pure evil. The transcendentals of the above Platonic trinity were considered wishful thinking, needless weights. All rules and dogmas were seemingly put aside by the paintings of Cézanne, the poetry of Baudelaire, and the architecture of Le Corbusier, the alleged forces generating modern art. Behind this proclaimed freedom Ellul found an absolute slavery, nonsense masquerading as sense. The denial of absolute limits was denied absolutely and became the modern aesthetic of *n'importe quoi, or* "*whatever,*" of "anything goes." An absolute denial of absolutes does not ring true. If all is possible, nothing is possible, which must become possible; these are major contradictions that belie the technological imperative, a necessity disguised in a vehemently asserted "free choice." This problem is central to Ellul's enterprise in *The Empire of Non-sense*.[10]

In the past "anything" was judged by whether the community profited, by whether the gods allowed what was proposed, and by whether the natural world could be so manipulated and arranged. Art on cave walls likely served magical powers, enabling the human with luck in the hunt; buildings were constructed to convey certain honors and sensibilities—for example the door of Notre Dame Cathedral was large enough for God to enter; paintings of the eighteenth century displayed what the middle class hoped to own and control. But, what does a red X drawn on a gallery wall portend?[11] Who or what is being honored, informed, or pleasured?

Modern art literally exploded with examples that seemed to challenge every approach to art: representational art gave way to abstract art; art with a message was reduced to formal considerations; engaged and political art becomes anti-art. These transformations occurred across the board in music, painting, sculpture, and theatre and were also tied to obvious technological developments. Abstract painting became more so with the developments of new pigments and surfaces, with microscopic accoutrements; music could be played on electronic and computer controlled instruments beyond human control; John Cage offered a greater challenge in a concert of sonorous silence. Developments in mathematics and computer applications played out in all of these areas. Sculpture went the same way. New materials and methods of calculation produced remarkably diverse possibilities. And, of course, architecture was the playground of technique. Buildings were possible as never before because of new materials and methods of application—new ways to heat and to cool, new landscapes to conquer because of technical advances, like Frank Lloyd Wright's house Fallingwater. In all cases boundaries were and are challenged. Ideas of harmony both in color and sound were challenged. Atonal music following mathematical formulae and ear-splitting screeches called into question all that had been considered acceptable. Free at last. But free from what? The answer: tradition, both cultural and material, if these are separable.

Then, poetry, the novel, and theatre offered different challenges. Images dominated the plastic arts, but the literary arts lived and died by the word. Like the plastic arts, the novel and poetry broke from their traditions most clearly in their attack on language, the material of their work. The novels of Robbe-Grillet denied metaphor and often narrative structure; the poetry of the Beats abandoned melody, rhythm, and rhyme for often inarticulate vocalization (from Ellul's point of view); plays were performed with no plot or script and the audience was asked to make up its own drama. Boundaries, again, were attacked. For modern art, anti-art is always around the corner in the challenge of limits—perspectives, standards, values, and meanings that transcend the act or example. The word, further, collapsed into the image, meaning into what was merely present and to a discourse banning opposition although open to simple negation.

Ellul claimed that art with a message and formalist art were the hot and cold taps of technological art, which generated apparent oppositions:

On one hand, we have art totally integrated into the technical system, an exact reflection of that technique in its frozen perfection, in its meaninglessness, in its neutrality, its indifference to pleasure, to beauty, to suffering, to the intolerable. On the other hand, we have an art that shouts out society's disorder in the grip of technique, an art which struggles in revolt against what it no longer knows and which questions everything simply because it is the thing to do. This art only expresses chaos under the impact of techniques that disintegrate and distort man's traditions and meanings. Such are, I believe, the two foundations of the two contradictory expressions of modern art and the reason why they are so much of a piece, so meaningful, so dramatic (but only dramatic, not dialectical!) *in their contradictions.*[12]

I can state simply that, for Ellul, a dialectic involved holding two contrary positions together in such a way that their opposition did not collapse but resulted in a new meaning that was the result of seeing them together. A "no" is no opposition. An opposition between mind and body, we could say, produced tools. An opposition between mind and body also produced Descartes and his philosophy. We could further say that the oppositions between tools, Descartes' philosophy and the Enlightenment belief in progress, and the appearance of the technical phenomenon produced Ellul's theory of technology and his library of writing. These would be examples of the dialectic. In Ellul's view the oppositions between art and anti-art produced blather, where one collapsed into the other, where all of the supposedly freely chosen ways of working art were merely the manipulations of technique and of saying "no" to whatever past came by. Tied, then, to an understanding of Ellul's notion of dialectic was his theory of the symbol.

The symbol was the attempt to say or to show what could not be said or seen but to do so in any case. The symbol was dialectical. The painting was not the beautiful but an attempt to present it in images, nonetheless; the word—poetry, religion, philosophy, for example—was an attempt to grasp in a present—a puff of wind or marks on a page—what was beyond that present or any present, to reach what that present meant. Symbols are oppositions opposing, the essence of which technique denies. The clash between world religions, philosophies, economic and political policies is the confrontation of symbolic structures. Ellul has made the confrontations of technology and culture his passions, his focusing of memory and its embodiments in culture. Technology is the attempt to be meaning itself, the truly ontologically ontic, *being itself*, in a phrase. More on these points later.

Modern art, he argued, possessed no style but was the hodgepodge, an anthill of movement and discourse going off in all directions.[13] It had no clear sense—meaning—and no true embodiment. It was a denial of otherness; in its denial it became otherness itself, the ephemeral in motion. Modern art produced objects that were anti-objects, concepts temporarily embodied, in other words, technical phenomena.

III

The understanding of the technical phenomenon in the mentality of *la technique*, explained in *La Technique ou l'enjeu du siècle* (1954) translated as *The Technological Society* (1965),[14] is the key to the loss of dialectic, to the degradation of the symbol, and to the merging of all elements of cultural making. Simply, all making entered the domain of the technical phenomenon that was caught in the technical system, which canceled transcendental acts or objects. The oppositions occurring in the system were only dramatic, only an aspect of the show. The meaningful, and the symbolic, as stated above, required an abiding tension between elements opposed but that did not collapse, with a meaning that appeared in their opposition and not in their denial.

Ellul defined *la technique* carefully in the "Note to the Reader." The definition cannot be studied enough and has produced much misunderstanding and confusion. The reason is simple. It took an entire book to explain it:

> *The term* technique, *as I use it, does not mean machines, technology, or this or that procedure for attaining an end. In our technological society,* technique *is the* totality of methods rationally arrived at

and having absolute efficiency *(for a given stage of development) in every field of human activity. Its characteristics are new; the technique of the present has no common measure with that of the past.*[15]

Technique involved method rationally employed. All cultures have methods: they use tools, their hands, their imaginations, etc. to extend human intention into the world, and in all cultures tools are used to effect this will and way. Ellul understood the use of tools as technical operations that extended from the body and then also from spirit or intention. The body was the initial form of otherness and opposition for the human and then nature was its second other and negation. In relation to these forms of otherness humanity exercised freedom in the overcoming of these provisional necessities and also by meaningfully maintaining these relations as limitations against which humans could work.[16] With technical operations otherness was maintained. Ellul noted that advances in technical operations were typically achieved by imitation and slow alteration if they occurred at all. Flight was not possible until birds were no longer imitated.[17]

The technical phenomenon resulted when consciousness entered and attempted to perfect the tool with a mathematics-like method. Chain saws replaced axes; autos replaced horses; computers and calculators replaced the abacus; and so on. In like manner, paint brushes, chisels, flutes, etc. became computer-generated data and digitization. Technical phenomena proliferated, and they did so with a certain logic. As they advanced, the othernesses of the body and nature abated. Technique became a full-blown intention (*une intention technique*)[18]—historically Ellul located this period after 1750—when there was no longer a clear distinction between the made and the not-made, when objectivity collapsed into technical subjectivity and when society became an "other."

In chapter two of *The Technological Society*, "The Characterology of Technique," seven primary characteristics were listed and discussed: rationality, artificiality, automatism, self-augmentation, monism, universalism, and autonomy. Rationality is central. In the following, I have amended the Wilkinson translation and added a phrase in brackets that Wilkinson had left out:

In technique, whatever its aspect of the domain in which it is applied, a rational process is present which tends to bring mechanics to bear on all that is spontaneous or irrational. This rationality, best exemplified in systematization, division of labor, creation of standards, production norms, and the like, involves two distinct phases: first, the use of "discourse" in every operation [under the two aspects this term can take (on the one hand, the intervention of intentional reflection, and, on the other hand, the intervention of means from one term to the other)]; this excludes spontaneity and personal creativity. Second, there is the reduction of method to its logical dimension alone. Every intervention of technique is, in effect, a reduction of facts, forces, phenomena, means, and instruments to the schema of logic.[19]

Rationality resulted in the creation of a concept, an abstraction, in which moments, objects, a series, etc. were reduced to a logical schema and a discourse. From the logical perspective A equaled A, could not be both A and not A, and was either A or not A. The proposition, the concept *per se*, was the embodied technical phenomenon. Discourse was essential in the movement of consciousness from an awareness to an awareness of an awareness, a further leveling of consciousness in which a sense of time and body were co-opted. Mechanical techniques began to increase exponentially after 1750, but it was throughout the nineteenth century that technique was "naturalized." Even Marx thought that progress was possible from it.[20]

With the universal application of the rational to experience, to transform what was into what should be, a process of rational ordering initiated an unchecked growth of technical phenomena. The artificial was the truth of rationally applied technique. The natural world was full of distinctions lacking essences. The distinction between the made and not made collapsed. Whipped cream came from cans. The label and the process—soon indistinguishable—defined the product. Choices opting for technique were made automatically and resulted in a geometrical rather than arithmetic progression. Ellul called this self-augmentation, where two becomes four and four becomes sixteen. All the many kinds of soap are samples of this characteristic. How many ways can emulsification be made to appear, where appearance seems to be the reality? An infinity of addition, n+1, is the answer. The technical phenomena appeared as a necessity, both moral and ontological, in the characteristic of Monism. Here Ellul announced: that which could be made would be made. Use was inseparable from being. What weapon made was not used? Techniques became universal—the so-called global village was simply *la technique* worldwide, where villages were on the way out. The last category, autonomy, occurred when technique became the sacred: it would be improper, even sacrilegious, not to use it, to do it. Ellul distinguished the holy from the sacred by locating the sacred as a really graven image.[21]

Epistemologically, there is a movement, I have argued, between these stages. At first subject and object are separated: there is something before me on which I make my move. With tool using the separation between subject and object remains. But with the application of a mathematics-like mentality in the pursuit of absolute efficiency, in the search for the absolute itself, the distinction collapses. The true, by definition, must be the rationally conceived, at which point, the real becomes the rational. This occurs in discourse as well. Ellul is clear in his definition of rationality—there is a movement between awareness and awareness of awareness that is mediated by discourse, in this case a rational concept. The first aspect of the encounter is forgotten in order to create the identity logic requires. A can never be exactly A: it is either to my left or to my right as I write it down with the one becoming two, or one A is either before or after an other A as I speak it or think it. As this type of logic is the basis of technical rationality, it is clear that both time and space are to be cancelled, for example in the digital watch and in the so-called but never experienced global village.

Time as progress, of course, is allowed and even required as the infinite series of possibility is advanced, with the paradox that if there is no boundary to the series, a contradiction results in trying to conclude the series. That is, the so-called goals of efficiency and the "one best way" that help to define technology are impossible. The best is either the next one or not any of the ones. If it is the former, the infinite is finite; if it is the latter, the infinite is an empty class concept. Ellul understood that this infinite advance required making technology both an infinite series and the infinite itself. Technology became the sacred. This happened to art as well with the notion of beauty: either there is beauty as the next object is produced or beauty is none of the objects, in fact no object at all, but discourse, *la technique* as a form or rationality, itself. There is nothing beyond the moments of it, either in fact or in understanding.[22] In *The Empire of Non-Sense*, Ellul wrote:

> *Under the influence of the technical phenomenon the object becomes the only important thing, all that matters. The search for meaning, for beauty, for communication, for values, for moral ideas, for metaphysical enquiry, prevents seeing the painting or hearing the music, because it is*

painting "in-itself" and music "in-itself" that is the important "thing," the painted object, and there is nothing else to search for behind it.[23]

And, the object has been subsumed under technical rationality. In this way art becomes non-sense. It has no sensual manifestation and, ultimately, it cannot be understood, for that would be to confine it. The technological society becomes a system when all of the phenomena are linked together. This was made possible by the computer and the internet and all those other advances that support them, advances that can be barely known or understood as a result of the linking that technology effects.

Put more concretely, the more the technical phenomenon is manifest, the more the art object— or any object—is led to the formal, the abstract, and the systemic. We seek meaning that is always a step away. Consider the meaning of this particular moment—one letter leads to a word and soon a sentence appears. And then a paragraph and a series of them and then, suddenly, we have a book or an essay. To say the meaning of anything is merely the next moment is to call the meaning of that moment to question. Meaning chases moments, which in turn beg other moments, *ad infinitum*.

Attempt to find meaning while watching a digital watch. To do so is to give the moments meaning. Meaning, however, cannot just be the next moment unless that moment is tied to a purpose beyond the clock like going to the store, brushing one's teeth, or getting married. The meaning of these clusters of moments assumes a shape beyond them and opens to a greater search: do I go to the store to provide food for my future wife, do I brush my teeth to prepare for a passionate kiss, do I get married to have children or to increase the pleasures of two people and of those around us? Do all of these questions lead to the ultimate meaning of life in the universe, which in turn seeks further meaning? Suddenly, we enter a domain inviting a Dante or a Rembrandt or a Beethoven who could give us sights, and sounds and metaphors worthy of the complexities beyond any meaning a clock could provide.

The digital watch no longer suggests the universe with its concentric circles in motion, a meaning going clearly beyond gears and armatures as did the traditional watch. The digital watch further reduces moments to electrical impulses and to greater and greater abstractions that demand a physicist's aid. Dante's Virgil would not be wanted or needed or noticed. Thus, meaning moves outside the succeeding moments to the very reason for collecting or noticing them. If meaning moves too far beyond them, however, the moments become a barren series of moments. We are now at home, or rather not at home, in the bad infinity.

IV

Ellul held that development in the arts mirrored that of *la technique*, and we could understand that in the following way: in the tradition of painting, painters ground pigment in oil each day before painting, employing clearly a technical operation. Painters had to apprentice to learn the art of making paint before the problems of applying it were faced. Metal or "mental" tubes appeared by the end of the nineteenth century, which altered painting forever by allowing uniform colors, an ease of storage, and a convenience at all levels—and certainly a move toward efficient ordering. Rembrandt had made his own paint, and his canvasses were unique from the first stroke; his genius, imagination, and perspective provided the rest. One can currently paint in pixels, which can involve genius and imagination, but it also involves the entirety of the technical system including the power companies, the communication and transportation systems, etc. The list would be endless. But, missing would be the touch of the individual and the basic level of material creation and perhaps the level of purpose and goal. When all is possible, little remains except "whatever." Rembrandt's goals are gone. Painters no longer live in a world overseen by an active and benevolent God, which, one could argue, like John Berger, influenced perspective on all manner of levels.[24]

In the tradition of painting, paintings referred: perhaps the painter of cave walls, as I noted earlier, wanted luck in the hunt; the painters of medieval art portrayed their veneration of Christ and the Gospel endlessly, perhaps the painter in the eighteenth century hoped to sell a painting of signs of wealth, a full table, a verdant landscape to a wealthy bourgeois patron. But when the painting is about painting itself—lines, squares, blotches—the question of reference arises. Clearly, the technical system and its move toward mathematical abstraction are suggested. The notion that there is no predominant style or perspective mirrors the absolute of absolute efficiency as the next "one best way," which is another version of "whatever." We have, as well, the new sacred, a purity of purity. Being is what is before one—past, present, and future combine.

All the arts, as do all the segments of society, tend to combine. The distinction between the plastic arts and the literary arts merges with the reduction of words—a meaning that surrounds, to an image—a meaning that is before one. Ellul wrote:

> *Thus visual reality is clearly non-contradictory. You can say that a piece of paper is both red and blue. But you cannot see it as both red and blue at the same time. It is either one or the other. The famous principle of non-contradiction is based on the visual experience of the world, just as the principle of identity is. Declaring that two opinions cannot both be true, when one denies what the other affirms, has to do with vision, which involves instantaneousness. But language involves duration. Consequently what is visual cannot be dialectical. Knowledge based on sight is of necessity linear and logical. Only thought based on language can be dialectical, taking into account contradictory aspects of reality, which are possible because they are located in time.*[25]

With the technical phenomenon, the result of a logical schema, objects and processes turn toward images. The tension between word and thing relaxes, and the word as the prime conveyor of the symbol is humiliated. With this humiliation is symbolic degradation. Ellul understood this degradation as "clichegenic" discourse. The cliché is the machine in its new suit. Technical subjectivity collapses into technical objectivity—the technical phenomenon—making the cliché the language of this collapse—the subject and the object become indistinguishable. Coke becomes the real thing, etc.[26]

The creation of the technical system involved the linking of techniques such that no one technique was the cause of any other. The system predominates, with the following effect:

> *Technique cannot be symbolized for three principal reasons. First, it has become the universal mediator, and because it is itself a means—in its capacity as means—it is not the object of symbolization, but rather it is also, by its power, outside of all other systems of mediation or symbolization. It is, in the second place, a producer of a communal sense. The communal act today no longer relies on the support of the symbolic but rather on a technical support (the play of media, for example). Simply, technique establishes a non-mediated—an immediate—relation with*

man, who, in the past felt a strong need to distance himself from nature but technique seems not to require such a distance. It seems to be the direct extension of the body. Who has not heard it said that the tool is merely an extension of the hand? Thus, we pass from an organic world, where symbolization was an adequate and coherent function in relation to the milieu, to a technical system where the creation of symbols has neither place nor sense. What symbols are necessary are produced out of technique itself. Television or advertising offer abundant symbols of technique but these come from the very working of technique itself. Therefore, the technical milieu is never understood because symbolization is excluded. And, from this fact, art, the foremost minion of symbolization, finds itself chaotic and torn between its "vocation" and that to which it can no longer aspire: an environment made up of discrete pieces belongs to structuralism but not to symbolization.[27]

Tools extend, as noted above, and thus they refer. Achilles' shield is both tool and powerful metaphor. But, at a remove from the natural and organic world, the technical phenomena link to each other, where each one is not simply a means but is at once a mediator, a means, and a mediated. The meaning of *Star Trek* is itself. The notion of *Star Trek* can suggest technological homelessness, but this is to go way beyond what was intended or at least exemplified.[28] Ellul turned *la technique* into a symbol and some of his readers did and do not follow, as one would expect.[29]

The viewer or listener, the audience in any case, before technological art is often at a loss, not sure what is meant, unsure of the proper response, unsure if they experience pleasure or pain, and thus the need for the critic. The critic is the technician of art—sometimes the artist him or herself. The critic becomes the broker of art, the guarantor that what seems like refuse from a dump is indeed art. Ellul noted: "First and foremost, the critic guarantees the durability and lasting value of the work in question. The critic is just another business agent whose job is to guarantee status."[30] Status and value in question cannot be allowed outside the technical system, but at the same time they must sanctify the system. Ellul refused to call them the holy and certainly not the Wholly Other, or even an other. For this reason among others noted, art and ideology are never far apart.
Ellul noted:

Further, a system so demanding and rigorous, so far beyond human experience, begs an ideology that hides the reality of the system, while at the same time lending support for it and for its consequences. Hence, the more the system is rigorous, rigid, and totalitarian—in a word inhuman—the more it must conceal the actual reality of the ideology with compensations that make the situation tolerable (the ultimate costs of the automobile—city crowding, all manner of noise and pollution that remade cities—are veiled and compensated for by the passion for speed, by the manifestation of social prestige, etc.). However, this ideology reveals, at the same time, man's essential denial of this reality. Man withdraws either from his own life or from his former life, or from politics, or even from an aspect of the technical system. There is a fundamental refusal that is expressed solely in an ideological way that plays itself out at this level. Here, compensation enters: one mounts an ideological attack on the system in question with the result that, so satisfied, one goes no further. The ideology is self-justifying to the precise degree that it raises questions, but this is a justification in the second degree to the extent that it empties out and exhausts the human capacities to resist the inhuman.[31]

Art has retired its functions of providing a communal response to help form and found and then perhaps to criticize, by going beyond whatever empire is in place. Art was a creator and midwife as Ellul noted above. It is no longer figurative, no longer refers to those elements of daily life and hope, beyond the current exhibitions of chaos and dissonance. The technological society is no real community in Ellul's account, but again, as noted above, Ellul offered hope. Against those skeptics who still doubt his case regarding the transformation of art into technique, and about the primacy of "whatever," Ellul suggests this experiment regarding the art of cuisine, which is one of the most basic communal arts:

> *... take a glass of paradichlorobenzene and add it to a large tube of neoprene glue, throw in a trickle of ascorbic acid to bring out the flavor. Reduce under a low flame. Then cut up in wide slices of expanded polystyrene, dip them in a sauce of oil drained from a car and serve hot. You can torture the ear and attack the eye, but you cannot force taste to accept anything, or "whatever." Such is the limit of reality. But, the more you approach the true, the more you reach the centers of abstraction, of reflection, of mediation that is mental, the more the possibility of lying, of doing ill, or of destroying increases. Truth and "spirit" do not defend themselves like a stomach in revolt, and this is the problem. One can break apart language. Clearly meaning and communication do not produce the immediate reaction that one would have to sulfuric acid. Modern painting, music and literature—because they obey the law that "everything is permitted"—are in the same order as my cooking recipe, and since the effect is only on the nerves, on the psyche, on the intellect, on the ethical and spiritual make up, all of which do not register on a seismograph, then there's no need to worry. It's a simple matter of adaptation.*[32]

Clearly, everything does not go.

Notes

1 *Jacques Ellul on Religion, Technology, and Politics: Conversations with Patrick Troude-Chastenet*, trans. Joan Mendès France (Atlanta: Scholars Press, 1998), 22. For a complete look at the Ellul corpus see: Joyce Main Hanks, *Jacques Ellul: A Comprehensive Bibliography* (Greenwich, Connecticut: JAI Press Inc., 1984); and Joyce Main Hanks, "Jacques Ellul: A Comprehensive Bibliography: Update, 1982-1985," *Research in Philosophy and Technology* 11 (Greenwich, Connecticut: Jai Press Inc, 1992), pp. 197-299. And, finally, Joyce Main Hanks, *Jacques Ellul: An Annotated Bibliography of Primary Works* (Stamford, Connecticut: JAI Press Inc., 2000).

2 *Conversation*, 52. Also, see *In Season, Out of Season: An Introduction to the Thought of Jacques Ellul: Based on Interviews with Madeleine Garrigou-Lagrange*, trans. Lani K. Niles (San Francisco: Harper and Row, 1982), 14.

3 Jacques Ellul, *Perspectives on Our Age: Jacques Ellul Speaks on his Life and Work*, trans. Joachim Neugroschel, ed. Willem H. Vanderberg (New York: The Seabury Press, 1981), 15. I discuss Ellul's intellectual development in my *Technique, Discourse, and Consciousness: An Introduction to the Philosophy of Jacques Ellul* (Bethlelem, PA: Lehigh University Press, 1991), 121-136, hereinafter referred to as *TDC*. One would do well to consult Andrew Goddard, *Living the Word, Resisting the World: The Life and Thought of Jacques Ellul* (London: Paternoster Theological Monographs, 2002) for details of Ellul's life. He notes that Ellul is not always consistent about dates. Goddard explores Ellul's Christian dimension quite well, and his discussion of Ellul's view of law is very useful. Ellul's theology is quite complicated, however, and those interested in this area would do well to consult the field, which is quite large. The bibliographical work of Joyce Main Hanks is essential in this regard.

4 Jacques Ellul, *L'Illusion politique: Essai* (Paris: Robert Laffont, 1965). Jacques Ellul, *The Political Illusion*, trans. Konrad Kellen (New York: Alfred A. Knopf, 1972), n.4. Hereinafter cited as *PI*.

5 *Ibid.*, 29.

6 *Ibid.*, 219-223.

7 *Ibid.*, 49-50.

8 *Ibid.*, 61.

9 Jacques Ellul, *L'empire du non-sens: L'art et la société technicienne*, (Paris: Presses Univeritaires de France, 1980), 285, (trans. Johnson and Lovekin), hereinafter cited as *L'empire*.

10 *L'empire*, 59-60. Here Ellul makes his first attempt at explaining "whatever," but the explanation moves throughout the text.

11 *Ibid.*, 259.

12 *Ibid.*, 50.

13 See especially *Ibid.*, 58-9. Also see my discussions of Ellul's dialectic in *TDC*, 89-96.

14 Jacques Ellul, *La Technique ou l'enjeu du siècle*, (Paris: Armand Colin, 1954); Jacques Ellul, *The Technological Society*, trans. John Wilkinson (New York: Alfred A. Knopf, 1965), hereinafter cited as *TS*.

15 *TS*, xxv.

16 *Ibid.*, xxxii-iii.

17 *Ibid.*, 19-22.

18 *Ibid.*, 47. *La Technique*, 44.

19 *TS.*, 78-79; *La Technique*, 73-73.

20 *Ibid.*, 54-5.

21 See my *TDC*, 152-187.

22 This is one version of the bad infinity that I discuss in *TDC*, 98-105.

23 *L'empire*, 92.

24 *Ways of Seeing* (London and New York: British Broadcasting Corporation and Penguin Books, 1977).

25 Jacques Ellul, *The Humiliation of the Word*, trans. Joyce Main Hanks (Grand Rapids, MI: Wm. B. Eerdmans Publishing Co., 1985), 11, no.3.

26 See my discussion of this in *TDC*, ch. 6 and in "Looking and Seeing: The Play of Image and Word—The Wager of Art in the Technological Society: A Revision", *Bulletin of Science, Technology, and Society* 32 (4) Fall, 2012, 273-286.

27 *L'empire*, 68-9.

28 In my "Technology and Culture, and the Problem of the Homelessness," *The Philosophical Forum,* XXIV Summer (1993), 363-374.

29 See my *TDC*, 29-64.

30 *L'empire*, 255-6.

31 *Ibid.*, 107-8.

32 *L'empire*, 233-4.

THE EMPIRE OF NON-SENSE
Jacques Ellul

INTRODUCTION

What intellectual has not wanted to write pages that would decisively explain art, this mysterious innovation of humanity? But, I do not have the knack for metaphysics, and I have never pretended to delve into humankind's innermost self. In a more modest undertaking, I will consider our age and modern art both in full crisis and in full bloom, superabundant and manifesting itself in a variety of aspects. Can one discern a "principle" [*raison*] for this profusion, this ambiguity? I perceive an unbelievable vitality, which, as soon as it blossoms, withers, apparently without fruit. Is it possible to find a deep underlying current without giving in to mere classification, to find a guiding thread amid this chaos that is both fecund and sterile with contradictory theses and meaningless non-statements? These questions arise as soon as we cast a glance or lend an ear, even distractedly, to this mad abandonment. Our Western society and our age are called to question by the art it produces. Responses to these questions are typically couched in commonplaces. Clichés everywhere abound: those of the "imaginary museum," those of the transformation of science into art with a new type of knowledge as the result, those of art made popular in a mass society by its mass media. By clichés I do not mean foolishness or error. Without doubt, the world of art is transformed by science, by the media, and by the "imaginary museums,"[1] but I do not wish to rehash the obvious, to merely repeat these conclusions while avoiding the uniqueness of present-day art. Each one of these claims contains an element of truth, although insufficient by itself, an insufficiency that increases to the degree that each one of these, apparently idealistic, is based frequently in a biased and materialistic philosophy. I will pursue a more limited project. We find ourselves in the presence of a world of unique artistic creation, and we need to ask: from what source does this art draw its specific uniqueness?
I advance my working hypothesis that art, like any other activity today, is situated in a technological society and in the technological system[2] that provide the new environment and explanation, the reality and even tragic truth, of modern art. In keeping with this hypothesis, I will not attempt an examination of art in all its dimensions or propose an account of art in relation to the human condition, or, finally, to define the essence of humanity through artistic activity. We do not need to consider the universal history of art, to seek out its mysterious sacred origins, or to take up the ceaseless search for this lost dimension. I will not enter the domain of the psychology of art in its pursuit of profound and secret motivations or take up the sociology of art. These theories, so often vented, so greatly and passionately discussed in a philosophy of art, will not detain us. We will take as our vantage point a brief moment in art's long history, that of the last century. We will adopt a single point of view, that of the relation between art and technology in this century.

We do not intend to take into account the interesting theories of Georgi Plekhanov and of György Lukács, for example. As true Marxists, it is a problem to know how to explain and to justify artistic activity proceeding from the theories of Marx, while remaining within this context. They have reestablished the dignity and freedom of artistic creation for Marxists within the limits of Marxist theory, but that has nothing to do with our undertaking. It is an issue, once again, of taking a look at the eternal nature of art, or its permanence, of its eternal as well as its historical significance. Beyond the debate on the relations between the forces of production and superstructure, one can consider that Marx effectively analyzed the relation between modern art and industry and, therefore, the situation of the nineteenth century. In the nineteenth century, he saw that the "history of industry and the objective reality that supports it are the open book of the essential powers of mankind," which have never been understood except as abstract essences, such as politics, art, and literature, etc. To put art in relation to industry is to circumscribe its reality and also the reality of mankind, and one immediately perceives to what degree art is in exact reflection of what industry and capitalism has made of man; art expresses human alienation, the division between subject and object and the producer and the product. Through art, what exists is a by-product of what is thought. Modern art is transformed by the law of capitalist production; it is an object manufactured by the producing subject and offered to the consuming subject. But it is more interesting that for the Marxist "art constitutes a *Techné* resting on the non-development or on the underdevelopment of technology and integrates the consumer's world of alienated ideology into an ideal world. Consequently, art can only be suppressed with the de-alienation and total development of the real praxis of workers who take possession of the world for themselves." Art must lose its essence to the benefit of *technique* (*cf.* Kostas Axelos, *Marx penseur de la technique*). Hence, with the present-day development of technology Marx questions the reality of art and its survival within our society.

I will situate myself somewhat within this point of view. I will not attempt to address the persistent problem of the relation between art and technique. It would be possible, of course, to examine the techniques of art or artistic discovery in the evolution of technique, the transformation of art by technical conditions. These matters raise questions to positions that consider art as either "eternal" or as subject to historical change. Here we are considering art only in the modern era and do not attempt to establish valid and fixed explanations for all forms of art for all time.

It will no longer be a question of examining art and technique in their raw and undefined states.[3] In reality, of course, we only consider modern techniques. And what concerns us is not so much the influence of one over the other but more the matter of how art has become what it is, and, above all, what it means. Art could basically be understood as a group of signs that refer to something, but to what do they point? One could speak of artistic freedom, of a flowering of the imagination, of an explosion in all directions of creative faculties, as well as of a kind of delirium, a destructive rage, etc., but these do not concern us here. Nor are we interested in distinguishing between that kind of art and what produced it. We will take art in its relation to the modern technical system, believing that it is there that one finds all the meaningful characteristics and values of modern art. Thus, on the one hand, we are faced with a description of the underlying relations of art (but not of art itself) and, on the other hand, a progressive illumination of a group of meanings that will transcend both art and technique.[4]

Nonetheless, it is still necessary to specify the terms of our investigation. I have spoken of the technical system and not of the technical society. It is a much overused and unenlightening cliché to claim that art depends on the society in which it develops or that art expresses that society of which it is a reflection.⁵ It is quite true that the evolution of art, for example, depends on the evolution of taste, on the progress of science and *technique*, on the new theories of figurative vision, and on the transformation of social relations themselves. But no one has been able to describe exactly the totality or the correlations among these four great axes of investigation. Of course, one may establish a relation between such an aspect of art and the global context in which it develops. And through art I am able to explicate, to clarify, the social phenomenon. In one of the most admirable formulations, of which he has the secret, Adorno writes: "The *unified* work of art is bourgeois. The *mechanical* work of art belongs to fascism, and the *fragmentary* work of art, in the stage of total negativity, aims at utopia." In one sense, then, it might appear much more interesting to study the relation of *technique* to society. This permits brilliant theories for attacking, for example, bourgeois society or for showing that art was in the nineteenth century the greatest protestation against that society, unless one seeks, on the other hand, to show that art is always art, at all times and in all places. For André Malraux art is a sort of gnosis that allows one to escape the provisional character of every age and the finality of death. Today, as always, art is a kind of religion. Moreover, it is important to understand this art in the totality of contemporary society. Who can doubt that this is bourgeois? In this society and indeed in the outlook of the bourgeoisie, the artistic and the intellectual superstructure takes the place left vacant by religion; it justifies the social order and the powers of the governing class. The more the bourgeois society becomes utilitarian, the more art and culture pretend to gratuitousness, the more mass oriented and generalized society becomes, the more art claims to be unique and personal. "The subjectivity of the superstructure replicates the frozen objectivity of the scientific infrastructure," as Bernard Charbonneau has said. As well one knows that the bourgeoisie is eclectic by nature of its very liberalism. "The uncertainty of the intellectual and artistic norms allows free spirits to search elsewhere for the satisfaction they no longer find in their society." But the bourgeoisie has elaborated an entire ideology of art with its compensating nature, its capacity for escape, and its power to coordinate all human activity on an ideal level. But, at the same time, art was relegated to the unreal world in its domain of non-serious things. It was simply an agreeable pleasure. It was an ornament pleasing to those who had the wherewithal and the superfluous wealth and luxury; here the artist becomes both the trained dog and the tamed lion, being even more despised to the degree he was covered with gold. The more irrelevant they made him, the more they pretended to fear him; he would show his teeth under the guise of questioning all that society was. One grew accustomed to the idea that technique commanded the forms of art while the artist was considered the representative of true human value. Art became functional because the bourgeoisie was functional, but it also became post-romantic, evanescent, *Kitsch*, because the bourgeoisie needed the perfume of incoherence and ethereal ambiances. "Art nouveau" is "an epiphenomenon of bourgeois reactionary tendencies despite the fact that it was, at its inception, decked out in socialist thought and in Saint-Simonian rêverie." (Robert Delevoy). The bourgeois has produced an aestheticism, it has disguised utility as caprice and has subordinated art to the necessity of a décor; it is the dominant class that wants to be both decorated and decorative while playing a role (in history) for which art alone provides the scenery. Only the useless was aesthetic. But what is important to understand is that art alone, during this time, continually filled that function in all its directions. I do not speak here of official art. That would be too easy. But the avant-garde, the revolutionary artists, were also eminently suited for the role assigned to them by the bourgeoisie. Manet, Cézanne, Van Gogh were perfect complements. And it was none other than the bourgeoisie who consecrated them! These artists gave the bourgeoisie a sense of the new and returned to them the creative force they originally held. Their revolutionary function or vocation was perfectly illusory. They revolutionized a style of painting while they merely reflected the era and thereby completed their mission. They were first and foremost individuals, and let us not forget that the bourgeoisie was individualistic and that it lionized the all-powerful personality. Napoleon, Goethe … but also Van Gogh, at a time when life had become humdrum, each was precisely the ideal individual. The bourgeoisie needed its tame artists to give pleasure, while at the same time needing the savage artists who lived out what the bourgeois ideology had been proclaiming for such a long time. The artist was obliged to play out the tragedy of the individual.

It is well known, even evident to the bourgeois society, that our art is linked to that society; even the "decline of art" may appear to be the expression of the decline of the bourgeoisie (and it matters little, as Theodor Adorno has shown, that the artist belongs specifically to such and such a class). One could say as much about any modern society.[6]

In socialist societies art is even more reactionary and perfectly stylized. The relation between art and contemporary society, whatever its economic structure and ideology, is clearly characterized by massification, standardization, reification—and art, when it wants to defend its integrity, gives birth to traits corresponding exactly to that society which it opposes. One continues to see under all regimes a "Camelot" corresponding to convenience—leisure, relaxation, conformism, digestion, consumption—and an "avant-garde" that is none other than the ideal aesthetic and formal projection of what is precisely absent. Adorno shows, for example, that music, both traditional and *avant-garde* remains essentially identical in its relation to modern society. When artistic modernism is thoroughly understood, the *avant-garde* always claims to be other than the norm, to which it poses fundamental questions, but when one examines the avant-garde of the previous century or of the beginning of the twentieth century, one quickly realizes one's error. They were confined to confirming or to concealing or to enabling or to compensating the existing social order that they claimed to put on trial. In this regard, the questions that Pierre Daix raises about modernism seem a little over-simplified. He remains convinced that modernism was truly explosive and liberating, but wouldn't it be better to say that it "closed" in on itself? Today, wouldn't being "modern" imply wanting to leave modernism behind? But, Daix says that modernism necessarily renews itself and, (renewed *hic et nunc*) becomes the authentic criterion assured of Truth and probably of Beauty. But one thing should put us on alert. When he speaks of what modernism opposes, he adds, "The exterior world, *in toto*." An important point: the exterior world against which the inner productive freedom of art rises up … This is a fundamentally bourgeois form of thought. Let us leave behind the dirty real world in order to go off into dreamland, utopia, and artifice … The Trip …. *dropping out*.

Of course, this is exactly their role. But Daix should reflect on this definitive formula: "Painter, don't worry about being modern; unfortunately, in whatever you do, you cannot avoid being modern." (Salvador Dalí). This is the final word on the value of modernism. It is nothing more than the expression of necessity, involuntary obedience, but worse, because it is involuntary obedience to the modernist imperative. Let me say once again that in writing this I do not pretend either to cast judgment on the aesthetic value of works but only on their claims to be innovative, revolutionary, and the embodiment of freedom or of whatever else. That is enough. I do not intend, therefore, to delve into the relation between art and society in general. Many others have done that. It is somewhat presumptuous, furthermore, for one to claim to confront directly this Brownian movement with its endless creation of new schools, which wave the banners of modernism, impressionism, cubism, abstract art, montage, *ready made* art, suprematism, surrealism, expressionism, mannerism, constructivism, neoplasticism, pointillism, abstractionism, informalism, found art, and so forth.

Let us limit ourselves to art and modern technology. But, first a question. Is it really a matter of what I call a technological society? Pierre Francastel denies that our society can really be technological. But his evaluation is based on a complete lack of acquaintance with modern technology and on a sociological analysis of our society ordered by some very clear ideological presuppositions. I will not rehash these problems that I have dealt with at length elsewhere. Our society is defined, determined, and characterized by its technology. But there are in this society many other things besides technology! I have never held to a narrow and unilateral determinism, nor to a mechanical view of cause and effect. The factors that combine with technology are many and diversify its expressions and products. Although it is absurd and unrealistic to say that technology is not the constitutive characteristic of the modern world but rather is "harnessed to serve certain economic and social goals," it is clear that modern society contains various forms of polarization. But what we are seeking here is the specification not of the relationship between art and society in all its diversity but rather of the technological system in its specific nature. Of course, we will need to digress continually and to consider as well, and secondarily, other factors that influence art, other roles played by art. But it seems to me necessary to tighten our terms of analysis in order to clarify this complexity. Even more than in other studies on *technique*, we follow Guy Debord in his remarkable insight in *The Society of the Spectacle*. Art is one of the fundamental aspects of a society of spectacle, being in part defined by technology. The spectacle characteristic of our world is not an automatic, quasi-natural product of the development of technology, but rather it is because that society is a society of spectacle that it chooses for itself its own technical content. And this fact allows art a new place with new forms. For if isolation is the foundation of *technique*, the technical process isolates in its turn, and art finds itself in the process of isolation. But it is clear that this study continues my research concerning technology and the technological system and is not just an illustration or an appendix. Art has its own specific nature. It remains, although inserted into a new ensemble of relations, a unique activity, the expression first and foremost of mankind's will, his quest to transcend himself, his most profound intuition. … And, what, then, does all that become in the world of technology? Also, what becomes of technology under the shock of this enormous project? Is art still what we believe it to be? And if it is undergoing a radical mutation, what ultimately does that mean for mankind and his destiny and for the image he creates for himself? That is our essential quest, which is neither metaphysical nor total. Concentrating on this reality that is expressed as such in our daily life as in the most sublime undertaking, could we have missed the mark?

* * *

Anyone can find a certain relationship between art and the new world that has developed along with the industrial era, and since art was profoundly influenced by it, one can easily find justifications for this fact. The most obvious step was to challenge on the theoretical level the contradiction between art and *technique* (but, of course, all the while disregarding what modern technology had become that was radically and essentially different from that which had been previously called technique!). Art itself is a technique.[7] Hence, art will be conceived both as technique and as an instrument for exploring and informing the universe. Art does not participate in the domain of the absolute but in that of the possible. We reject its essentially symbolic character (does theory appear reactionary?) because we need to examine to what degree the technical milieu tolerates the symbol. These clichés (which underlie Francastel's entire book) attempt to justify in advance the technical world: one must be able to prove that modern art is the same as art always has been and that there has been no rupture, no rejection. From this standpoint *technique* is perfectly justified and nothing should question its magnificent ascent (and especially not art). And, finally, there is an extraordinary alliance between the enterprise of modern art and the discoveries of science and technology. If the art of our time does not reach its full development, it is only "because a bad set of theories inhabits its implementation in the practices of the present-day world." We approach this study with the presupposition that technology is the only serious reality; that it represents today, as it has for some time, the only positive value, and that it generates all current human values. In this undertaking to justify modern technology, art, seen as a tool of justification, only raises the problem of knowing where to situate the stable relations between progressive techniques in the modern world and the strictly imagistic style of our age. Let the reader be pleased or not. Of course, from this perspective, all theories that contest or question this glorious progress, for example Sigfried Giedion or Lewis Mumford, are considered reactionary. However, this perfect marriage between art and technique raises small problems for some and we, like others, are faced with the need to provide art with a new "definition," or else to assign it new ends. If we insist on its fictive or symbolic character, there is an embarrassment *vis-à-vis* the supremacy, the totality, the efficiency of technical productivity, which apparently ridicule the fictive and the aesthetic and leave no room for the realization of an immaterial aesthetic content. First, we will recall that obvious truth that art is always artificial, a construction, a synthetic production. It manufactures an artificial world. But, in these conditions, how could there be a conflict with technique that does the same thing … etc.?

We see the implications.

More important is the assertion that art is nothing more than a game. Quite often we will need to see that art, in effect, is considered as a game. It is reduced to this by *technique*. (And this perpetuates the bourgeois concept of art as a superfluous leisure-time activity.) But saying that is to admit that perhaps art has declined and has been marginalized. But, on the contrary, we would be assured by bourgeois complacency that art always has been a game and never anything but a game. "Art began as a game, was pursued as a profession, and flowered as a cult—will it revert to its status as a game?" (Abraham Moles). Likewise, James Johnson Sweeny assures us that for art to keep its youth, art needs to play. The child plays at what he is not. He plays at sharing existence, "playing with, he plays at …." Such is art. All is art. And Sweeny seeks patents of nobility for his theory by recalling that for Schiller, also, in his *Letters on Aesthetic Education*, art is only a game and man is fully himself only when he plays. One could also recall *Homo Ludens*.[1] But it is quite easy to rehash all of this nowadays. It is a procedure of justification. We are very far from the profound and tragic considerations of Malraux on art. No, man was not searching for a meaning in life, nor for a sense of immortality, nor for influence on a mysterious universe when he painted the cave walls of Altamira. He was playing hopscotch and hide-and-seek. All discourses on the religious power of art are window dressings and afterthoughts. And, hallelujah, what luck. Thanks to runaway technicization, we discover the true meaning of art that so many centuries had obscured: a game, nothing but a game. But under these conditions, we

---

1 Johan Huizinga, *Homo Ludens*, 1938. [Trans.]

must challenge what we had invested in this activity. For a very long time, it was believed that art had something to do with beauty and that that was the recurrent concern—what constituted beauty and met its criterion? That was a mistake. Art has nothing to do with beauty. Those who speak of beauty and art are retrograde. And, if one simply wants to understand modern art or to situate it, one must renounce this absolutely irrelevant criterion. Harmony, balance of forms, the strict adherence to a framework or usage, grace or plenitude, whatever the criteria, it is always a question of beauty, but that era is closed. Baudelaire saw perfectly the antinomy between industrial development and beauty. "One can reasonably hold that a people, whose eyes get used to considering the results of materialistic science as the fruits of beauty, has, in time, uniquely lost its feeling for the ethereal and immaterial …." It is not for this opinion alone that we offer this citation (how many other authors in the mid-nineteenth century wrote the same thing!) but rather because Daix takes it up to show to what degree Baudelaire was mistaken. But he has completely missed the point. By using the criterion of beauty, Daix cuts off all possibility of evolution in art and of its very fertile relation with the technical universe. To the degree that one is concerned with beauty, one disregards the real direction of modern art. If beauty is defined arbitrarily, for example as a "power of revelation" (Gaëtan Picon), then everything and anything whatever *(n'importe quoi)* is permitted, and the term loses all meaning. Others have likewise attempted to preserve the term beauty by giving it a content suitable to what art has become in the technical world: "What is sufficiently structured," "What has enough of its own substance to impose itself" (they do not say on what …). But these are the pious formulas intended to save this traditional term from shipwreck, but we prefer, ever so strongly—those of us with courage—to say that we are indifferent to beauty and that all that is produced today as art has nothing to do with beauty. And so at the threshold let us abandon our ideas about beauty and our emotional attachments to the beautiful and the search for formal and aesthetic ideals, all of which prevent us from being objective and from seeing clearly that the art of our time has no interest in the transcendental domain. Perhaps we should ask why.

Likewise, one needs to abandon another commonplace about art. Too often art is conceived as what is given us for pleasure, an inner blossoming, an exaltation on a path of exquisite delights, heightened and rapturous joys of self transcendence toward something greater. On the contrary, this is a mistaken view of art. It is no longer a means of procuring any sort of pleasure for the listener, the viewer, or the reciter. Pleasure? What banality! It is, one could say, the culinary side of art! The element of gustatory pleasure is no longer compatible with the elevation or power, with the intellectual or spiritual dimensions of art. Modern art is meant to plunge us into horror, stupor, folly, release, ignominy, filth, sadism, masochism, all that you would want. Above all, there is no joy, no pleasure, no sublime, but just an abyss. "Music is no longer defined as a combination of pleasant sounds to the ear (according to Jean-Jacques Rousseau) but rather as a perceptible structuring of isolated and recognizable elements of sound (Moles) and here is one side of modern music: a cold algebraic blend. But, on the other side, is the polytonality of Iannis Xenakis, which is an audiovisual cataclysm … where low throbbings and strong resonances, light hints of percussion, which soon join forces and become as cutting as a steel wire, screaming out and piercing like an icy rain." So much for the chilling aspect (and let us make clear that these lines are not at all critical, coming from an admirer. I could accumulate hundreds of texts on the challenge of joys and pleasures in art. Here, one must again ask why. In short, one must abandon all that tradition considered as art in terms of content and goal. The art of today is perfectly artificial, no longer referring to nature and not produced to "raise" us up to anything. One could say, along with Vasarely: "Painting is dead." Romanticism, as well as classicism, is dead. Sense is dead. The naturalistic representation of the world is dead, and beauty no longer interests us. One triumphs when the questioning of art reflects its traditional function. One is carried away by the sentiment of mockery, which leads to a questioning of the "legitimacy of art," (but when one questions in this way we have an indication that art no longer exists). Art is only a game, on the one hand, but on the other (Veneigem), we learn that art is no longer the refuge of the gratuitous act. Art is no longer and in any way an autonomous domain, a garden reserved for delicacies and delights. Art plunges us into "real life." (But what is that?) Art no longer exists as art.[8] We are witnessing a permanent denial, a continuing production of anti-painting, anti-literature, or anti-theater; even theater masquerades as anti-theater. (Here again, one must ask, why this insistence on calling what one does anti-art.)

*　*　*

We are not really, in our investigation, hoping to illuminate the eternal nature of art or to unveil a transcendental aesthetic or to explain what motivates man when he paints or poeticizes. More modestly, in the here and now, we observe a certain quality in modern art. We see that it is situated in a new environment—the technical environment—and we notice, finally that it is quite radically different from all that which, until now, has been attempted in producing a work of art and all that which has been called "art." Of course, I know that many will object, but in each age, every innovation has produced turmoil, and one has always considered new art as scandalous, having nothing in common with what preceded it. The artists of the Renaissance deliberately broke with Medieval art; the Byzantines broke with Greek art, Hellenistic, and Roman art; and the Romantics broke with neo-classicism, etc. In each instance, one saw the same break deriving from innovation. But I believe (and I think I can prove it), that these breaks had nothing in common with what we are now witnessing, because the environment in which these changes occurred remained essentially the same. The innovations most often came about because of "something new to say," as Erich Auerbach has convincingly demonstrated in *Mimesis*. Art changed as a result of a new concept of life, a new scale of values, a "philosophy" or a religion that was changing. What was to be modified was the expression. But we are changing registers when people proclaim that essentially there is nothing more to say, nothing more to express and that there is no content and nothing signified. Yet many tried to maintain continuity by arguing that actually a change of philosophy has provoked modern art, and that now, science teaches us to see the world in a different light. The experts in aesthetics, by explaining that science teaches us to see matter and color differently, attempt to prove that nothing has changed, that the same old process continues. I believe that this is mistaken: the new expression of aesthetics does not come from the old process but comes instead from *technique*. The essential difference lies in the fact that when Renaissance artists wrote or painted, they attempted a new concept of man; the Byzantines, likewise, attempted a new theology, and so forth. Modern artists don't seek to express science. In other words, two radical and unprecedented changes have occurred: the milieu in which art is placed and the processes that evoke change impact art uniquely. Technical innovation has no common measure with what has gone before. Through these differences, the creation of the new, we approach the more essential question of the change in man himself. For if, as we observe, an activity as essential as artistic production is so completely altered, then humanity must be likewise altered. We will try to find a meaning, if there is one, for that strange evolution, which produces an art that plunges into chaos and yet is rigidly frozen as never before.

A Partial Bibliography

I am indicating here the principal books that have served me for producing the text. Within the chapters I will footnote specialized books on each issue I address.

Adorno, Theodor; *Philosophie de la nouvelle musique*, éd. Franc, 1962. *Einleitung in die Musiksoziologie* (1962).
Attali, Jacques; *Bruits*, 1977; *Noise: The Political Economy of Music*, 1985.
Cazeneuve, Jean; *La Télévision*, 1976.
Charbonneau, Bernard; *Le paradoxe de la culture*, 1965. *Le système et le chaos*, 1975.
Clair, Jean; *L'art en France. Une nouvelle génération*, 1972.
Combet, Georges; *Calliope et Minos*, 1972.
Daix, Pierre; *L'aveuglement devant la peinture*, 1971.
Debord, Guy; *Society of the Spectacle*, 1995.
Delevoy, Robert-Léon; *Dimensions of the Twentieth Century*, 1965.
Dufrenne, Mikel; *L'art de masse n'existe pas*, 1974.
Francastel, Pierre; *Art et technique*. 1956.
Giedion, Sigfried; *Mechanization Takes Command*. 1948.
Goux, Jean-Joseph; *Les iconoclastes*, 1978.
Habermas, Jürgen; "Technology and Science as Ideology," in *Toward a Rational Society*, 81-122.
Huyghe, René et Rudel Jean; *L'art et le monde moderne, I. II: De 1920 à nos jours*, 1970.
Leymarie, Jean; *L'art dans la société d'aujourd'hui*, 1976.
Maldiney, Henri; *Regard, parole, espace*, 1973.
Mantovano, G.; *L'arte nella società contemporanea, Studium*, 1969.
Moles, Abraham; *Art et ordinateur*, 1971.
Mumford, Lewis; *Art and Technics*, 1952: *Myth of the Machine*, 1967.
Nir, Yeshayahu; *Télévision, contraintes et perspectives*, 1976.
Onimus, Jean; *Réflexions sur l'art actuel*, 1964.
Read, Herbert; *Art and Society*, 1945.
Rookmaaker, Hans R.; *Modern Art and the Death of a Culture*, 1994.
Schaeffer, Pierre; *Pouvoir et communication*, 1972.

Notes

1 Ellul refers to the term used by André Malraux in *The Voices of Silence* referring to the effects of museums on European art. Trans.

2 The technical system, which is the product of the wager of *technique* and which constitutes itself as an autonomous and self-reproducing organization in the strict sense of the word "system," must clearly be distinguished from technical society in which it is situated and organized and by dint of which it develops. But it is evident that society contains something other than mere technique, and even though it is deeply influenced by *technique*, it also contains traditional traits and irrational drives and impulses. On all of this see, Jacques Ellul, *The Technological System*, trans. Joachim Neugroschel (New York: Continuum, 1980).

3 I will not attempt to repeat the remarkable work of Georges Combet, *Calliope et Minos*, 1972, on Art and Technique, because there, he examines the constant relation between the two.

4 I locate myself, evidently, in opposition to the theses of H. R. Rookmaaker (*L'art moderne et la mort d'une culture*, 1975): this work is very interesting and well researched but remains perfectly idealistic and considers that the evolution of art begins from the fact of the invention of new principles and in relation to new conceptions of life and philosophies. The rise of modern art is presented as a "battle" with a "victory" for its creators ... as if they have done something beside expressing the dominant sociological current! Art always remains, in their eyes, a search for meaning in life. This is very characteristic of classical thought on art.

5 Here, with his typical radicalism, P. Bourdieu (*Les fonctions de l'art—L'amour de l'art*, 1966) contends: "The principal function of art is of the social order ... Cultural practice serves to differentiate the classes and subclasses to justify the domination of one over the other." This leads us nowhere.

6 The distinction established between bourgeois art and proletarian art is nothing other than the product of a conformist conception and a repetition of a pseudo-marxist common place (for example, M. Ragon, *L'Art, pour quoi faire?*, 1971). Whether art is bourgeois, Christian, proletarian, etc., it is, first of all, included in the technical world, and it is there that it receives all its characteristics.

7 We will leave aside the triumphant discourse of innumerable authors on art and technique who base their thought on the decisive and brilliant observation that, in Greek, *Techné* means Art. And from that everything is resolved! A most profound thought! These small etymological exercises are perfectly ridiculous. I would limit myself, for example, to emphasizing that art does not mean technique in the etymological sense. Indeed the etymology of art is A P Ω (hence *ar-mus, ar-tus*). The Greek root is A P producing *artuô* (united by linking together)—aretê (suitability, excellence, perfection, and later on, virtue). Literally, from the etymological point of view, *art* signifies what is arranged, what is closely connected, and what is harmonized, and any positive and active capability. In other words, when one triumphantly proposes *Techné* as Art, one has said absolutely nothing at all, since the converse proposition is false!

8 It is definitely the case that art in its traditional sense is strictly handicapped. "One gets the impression that art is essentially superfluous after having seen the Dokumenta Exposition at Kassel (1968). After having strolled through the section on Op art, I realized that Piccadilly Circus or Broadway with their neon signs are much more fascinating, entertaining, richer in meaning—and, perhaps, more 'artistic.'" Rookmaaker.

Chapter 1
The Contradiction

We must start from the very clear idea that art veils society and expresses the deep reality in which we live. Virgil Gheorghiu's *The Twenty-Fifth Hour* ends with an appeal to the poet who reveals the dangers and the vices of society. In his work the poet expresses consciously and directly these dangers as contents of his work. However, we have gone beyond this stage with a veiling produced in completely unconscious ways. The poet, through the very style of his poetry or in his denial of poetry, speaks unintentionally from and of the depths of his age. Jacques Belmans, in his book *Cinéma et violence* (1973), effectively shows how films are the "implacable revelation of our public and private behavior." The least evident of these indications are the most important: social and psychological violence. Film does not limit itself to showing social and psychological violence, which it translates in depth and unconsciously. It holds up a mirror to us. Likewise, and more subtly, Robert Delevoy,[1] who believes that art is an establishing of order, states this essential idea: "Whether acoustic, visual, or verbal, the contemporary image could, by itself, present an order that stands over and against the chaotic state of the universe." New art might be a search for information in the struggle against entropy and disorder that it could express in an analogic, all-encompassing, and intuitive form of thought. But, Delevoy fails to discern the nature of this disorder and the type of society in which it occurs. More often than not, he sticks to a very traditional vision of a type of social or economic disorder. Nevertheless, his adjectives—analog, all-encompassing, and intuitive—are highly revealing against a world presumed to be analytical, rational and realistic. We ask: are his qualifiers more essential than those usually repeated in leftist circles? Delevoy also observes a rupture in art and in "taste." He has taken this idea from photography: the photographic image imposes the idea of an "objective" reality *vis à vis* the presumed illusive, imaginary and interior realities conveyed by the artist. "It happens that for the first time in history two parallel types of rival and antithetical forms share the allegiance or the resistance of the public. Taste is divided." Delevoy here grasps the profound reality of modern art. But the division is much deeper than he imagines. Our project is to show that modern art reflects and reveals technical society and is, thereby, fundamentally torn apart in all of its tendencies, which is a characteristic of the society itself. Even in its opposition to society, for example in its refusal to be technical, in its longing for the unrefined, for the material, for the spontaneous, for the unelaborated, for anti-art, etc., art exactly reflects this society. This art of refusal does not escape the process of reification; in its efforts for integrity it mimics those characteristics of the society it claims to oppose. The appeal to the "avant-garde," to what has come before, whether it be afro-primitivism or *Kitsch*, to what had been abandoned or fallen into decay, goes hand in glove with the denying tendencies of the age more clearly than art that is openly destructive. "This order that proclaims itself is nothing more than the mask of chaos" (Theodor Adorno). Such is our situation. Guy Debord carries this analysis further: "Art in the time of its dissolution, existing as a negative movement that pursues its own self-transcendence in a society at a specific moment in history where history is no longer experienced, is both art for the sake of change and a denial of the possibility of change. The more grandiose art's demands, the less they are realized. This type of art is, by nature, both *avant-garde* and not *avant-garde*. Its status as *avant-garde* resides in its dissolution.[2] This describes perfectly our experience in the world of art for a quarter of a century and what Adorno has already described as "the eclecticism of destruction." But we should be careful: this negation, this rupture, this destruction, does not express a society that is itself broken or dislocated. On the contrary, this society is rigid, homogenized, centralized, univocal, efficient, and rational … Adorno exposes the most profound crisis of this society: a powerlessness embedded rigorously in the organization and the irrationality of the technical system. Art plays its role in the game. By expressing, revealing, and illuminating the contradictions in the society, it reinforces and crystallizes them. Anti-art produces the same irrationalities and anti-social behavior by uncovering these irrationalities. It produces explicit and visible models of the collective and obsessive fantasies of death, meaninglessness, and destruction. Modern art is the art of self-negation and of a powerlessness to dominate the situation. It is the inverse of magical or religious art that gave man a force and a power to order the world. Modern art prepares the sheep for the slaughter in the continual declaration that there is nothing to be done. Modern art is the random agitations of false movements mimicking the mad buzzing of flies in a bottle. Openly challenging the tendencies of the technical world, modern art provides man with all the reasons—the rationalizations, the excuses and the flattery—to be subordinate, silent, and passive within the chains of the system. "The industrial administration of our cultural heritage has extended its power in the realm of aesthetic opposition." Technical display has its function. One need only look at present-day advertising images. They absorb everything. 1968 was a good year for advertising. Anti-bureaucratic critique became an excellent demonstration for office furniture.

Art joins forces with an imperative to order. The crucial moment in this process of absorption and reutilization is where art claims to be dominant and free. Bernard Dort (*Théâtre réel. Essai de critique* (1971) points out this involuntary copout. For him, modern theater is an introduction for the real. The praxis of the theater is essentially didactic and dialectical. It expresses, thus, a kind of revolutionary culture, but the revolution in question continues to be fixed on the Brechtian model, which is to say in a Marxist, anti-bourgeois mode, and Dort is simply mistaken about the type of reality and revolution to be expressed. He starts from the theater that was revolutionary in 1925, and he presumes this analysis to apply and to provide a criterion for all other types of theater. But here, with a lack of clear thinking, one is trapped. Nothing today is more conservative than the theater of Brecht, precisely because it attempted to provide a response, a model. But, unfortunately, it is a response to a question that is no longer asked with terms that have vanished in the roar of jet engines … Music, today, is called into question to the degree it considers itself free: artists have believed that the anarchic state of music or painting constituted liberty, but what exactly does liberty mean? It means that this so-called liberty produces the very image of the world revolted against. Adorno has said it best: "Music forges ahead within a false order … it marks the decline of art into a false order. … For as long as art, which has been made according to the categories of mass production, goes along with ideology and as long as its technique is a technique of repression, then non-functional art obtains its function. By itself, in its most mature and meaningful output, non-functional art projects the image of total repression and not its ideology." The crucial point in this subordination of art to technique—of art's impotence—or in its need to incorporate with the "irreconcilable" nature of the technical system, is the moment when artists pretend to be fully independent of *technique*, as if it were a simple banal instrument. When one says that the artist is the demiurge manipulating techniques, creating a new ideal of art (spontaneous) to which techniques are subordinate, (the basic position of Pierre Francastel—and whether the motive of domination is spiritual, philosophical, or political matters little—we are in the moment when art, or creativity, and invention are radically and totally incorporated into the technical system. This announcement is a lowering of the ideological veil (already noted in the works of Dort) that man throws over his own activity in order to hide its true meaning. The possibility that art no longer has any meaning and value, any true role in human life, would be intolerable and an admission of a decisive setback. None desire this: neither the technician, nor the bourgeois, nor the revolutionary, nor the writer, nor the artist, nor the public. Everyone is an accomplice in rejecting this integration of art into the technical system, but unknowingly or not, the production of art is incorporated nonetheless, behind the ideological veil, through its desperate insignificance, its marginalization in silent suffering. (An open suffering would be precisely that which is not suffered. An anti-imperialist theater points to the uselessness of explaining the politics of imperialism.) This modern art, salvaged for the sole purpose of being salvaged, expresses the fundamental reality of our time, which also maintains itself within the authentic line of art. Where one rehashes the evidence according to which art and technique conjoin (*Techné*, etc.), one must not forget Jürgen Habermas's cogent demonstration of the decisive contradiction between the two, and this is particularly relevant starting from the point when it is no longer a question of technique in general but of technique in our own time. Whether technique is instrumental and rational in relation to an immediate objective, whether it implies conditional expectations, whether it rests on the accumulation of knowledge and skill, whether it functions through solutions directed at problems, and whether, finally, it translates into a growth of productive forces, all this is exactly the inverse of art, which one can categorize among actions mediated by "symbols," referring back to an experienced world. The power to dispose of things technically is not at all of the same nature as the concept that social groups form of themselves to which, above all else, artistic activity contributes. This activity belongs to the domain Habermas calls *interaction* (but it seems to me that *Interaktion* in German does not have the same meaning in French, being more qualitative) and designates true practice (unlike technique) and assumes a symbolic representation while it mediates reciprocally between subject and object. Reciprocity leads to self-reflection; the consciousness that I have of myself is the result of a merging of perspectives, which has always been one of the preeminent functions of artistic activity that, today, is completely relegated to and negated by technical development. Thus, modern art, having been assimilated, remains, nevertheless, the desperate protest for a continuance of art *vis à vis technique*. This explains its double orientation.[3] First, we witness a differentiation between those who claim to be on the outside and attempt to take account of the technological society, those who expound on this society (for example, *L'Imprécateur* by René-Victor Pilhes, and also, *Les Choses* by Georges Perec), and those who plunge into the interior, who neither question the society nor the relation between man and *technique* but who translate this universal form of *technique* into art like the school of *tel quel*. The technical system speaks through their very mouths:

they are the *persona* that gives it both a face and a voice. It is not a question of being more or less integrated, more or less in the technical system: it is really a matter of two orientations, of two different outlooks, of two decisions qualitatively different. Technical society, on the one hand, and the technical system, on the other, put the artist against the wall. Inside these two tendencies, a movement of forms and directions flourishes, which does not appear as alien to our system. I could interject here, as a kind or preliminary to a step by step demonstration that: *the more rigid a society becomes structurally, univocal, and one sided, the more insignificant and self contradictory, diverse, incoherent, and inhospitable will be its ideology, at which point aesthetic décor will become a decoration in perpetual motion.* The result is a mobility of styles and modes, a mobility of works (Alexander Calder, etc.) and even a mobility in architecture (modular structures), to the use of aesthetics in what had been considered mere decoration (display windows, interior design, billboard advertising) and to the negation of the work of art as such (works of art that continually change or to anti-art that openly denies the reality of traditional art and its works). All of this shows clearly the predicament of art in the technical system.[4] Then, if we have understood the total contradiction between art and modern *technique*, as well as the subordination of art, the technical system that tears, fragments, and scatters art (and the society that produces it) the question heard for the last ten years must be repeated: is art dead? The greatest misperception comes to play here: if we mean that we witness the end of a formal activity that came to fruition during the nineteenth century and that abandoned a profound sense of human reality to conjure up theoretical notions of Beauty and to produce fantasies and dreams that extend a dominating, imposing, and unimaginative society, then we could claim that art is dead. Unfortunately, this seems the very function of art previously described (having been transposed of course!) that underlies the artistic production of our century. By contrast, we could claim an end to art as a relation to an eternal and transcendent quality in a symbolic expression of a felt and intuited absolute appearing in flashes of inspiration and a clarified obscurity. And if this were the case, the death of art, if it takes place, would also mean the end of man. Of this we can be certain.

∗ ∗ ∗

Modern art wants to pose fundamental questions but wants to ask them in a way that is too theoretical, too explicit, too definitive. This type of art, first of all, claims to connect life (directly and totally) with the ambiguity of this assertion. For, on the one hand, art is a matter of taking into account experience in the most direct, brutal, and least refined way. The outcry will become a privileged form of expression or else a direct attack on a sector of the public in whatever form it might assume. Music, painting, theater deny the separation of art and life.[5] This made John Cage's claim the guiding principle of happenings, the specialty of hippies. This has resulted in a mixture of politics, sexuality, play acting that challenge all kinds of constraint that include art and symbolization (as if there could be art without constraint!); and yet we also hear that one must make life itself a work of art. … Thus, the work of art and its productions are challenged. The world of objects is denied at the same time meaning is denied. We are told that we must consider the canvas as an object, as the painting in itself, without seeking anything beyond it, since only "the relationship between the subject and the codification of its system is meaningful" (Francastel), to claim that one of the great advances in painting occurs when it no longer seeks a fiction (perspective) but paints on a flat canvas that is itself an element of the work and, as a canvas, is the primary object.[6]

The work as an object is rejected and denied; the artist identifies with his art. He is himself a work of art. Long ago Oscar Wilde claimed to make his life an aesthetic work. To achieve this goal, one needs to act outrageously. Humor is not enough. The means are hopelessly the same: alcohol, drugs, sexual inversion, aestheticism, sadism, and at the extreme limits, murder as a fine art. Realistically, we see here a "confusion of genre": the project of making a life successful is a moralist's project. The mistake is to claim that life can be a work of art. Aesthetic criteria are inapplicable (except for self-contemplation, self-satisfaction; the claim of life as a work of art has never been more than the parade of Narcissus); to apply Epicurus's thought is not an aesthetic creation. The passage from an aesthetic object to the consideration of one's self as an object is a final pirouette to avoid the problem, a counterfeit profundity and engagement. Anyone who affects this attitude quite simply ceases to be a creator of art. He can convey an example of life (in his capacity as a moralist!), but he conveys absolutely nothing to future generations as a "cultural legacy." I know quite well that many will shrug their shoulders in the face of such an assertion but this is nothing more than a denial of art itself. The adventure of the Beat Generation poets is quite relevant here: Allen Ginsberg, Jack Kerouac, William Burroughs … poets "whose words made everything explode. The limits of the word, of life, of thought …." Poetry was only an outlet that expressed the pain of living, and they undertook to live this pain. They created for themselves a wisdom. They lived in the slums. They were wandering protesters. They considered themselves saints, prophets, etc. And, clearly, their poetry is a small fragment of their life, with both closely intertwined. But, when one reads these works without a partisan bias, one is dazzled by their poverty of expression, by their childishness … No, this is not *The Drunken Boat*. It is a series of blubberings and self-absorption. I am a fascinating personality. Take a close look at me. I'm going to reveal myself. André Gide's rheumatism does not interest me. I don't see why I should be interested in the blisters on Kerouac's foot. The life of Kerouac as a vagabond intellectual is interesting in itself but its artistic expression is little more than a trite monologue or else the indirect description of the Beat micro-environment. But, in the end, this expresses the other side of the claim for a connection between art and life.

Connect up with life! The rallying cry of the surrealists, "to smash the dichotomy created by bourgeois culture between art and existence" (but soon the author is talking about breaking out of the normal system of perception. Experimentation with drugs and hallucinogenics is a method that is both heuristic and subversive … this, in effect, is the common thread of the Beats. But is it life, existence?). Robert Rauschenberg said at the same time: "I am for art but for art that has nothing to do with art. Art has everything to do with life but nothing to do with art." In all these exalted proclamations, it is embarrassing that one has absolutely no idea what they mean by "life" or "existence" etc.

Strangely enough, those who have attempted to make their life into a work of art are led to self-destruction. Watts, who explains time and again that LSD broadens our consciousness, sings the same tune: we are only a part of nature; I am no longer limited by my sack of skin, which means that I am diffused into the universe. He sees one's self as the center of decision-making and consciousness, as a pure and simple convention, an arbitrary framework. By all means. But, then, the logical conclusion is suicide, since nothing makes me different from the rest, from little atoms fused with others. And, indeed, the behavior of all of these theoreticians was suicidal. The following contradiction resides in all of this art: the aspiration of an aggressive and hyper-developed ego ends with the abrogation of the life instinct.

Likewise, the art that is so tied to the technical society is, simultaneously, an art of *escape*. The absorption of "primitive" arts, the influence of the *Musée imaginaire* of the third world, is an expression of that absolute escape acted out by all present-day artists, an escape toward origins. Paul Cézanne declaring, "I want to paint the virginity of the world" is the prelude. An escape as an originating act is the most ancient gesture of man: Tal-Coat is characteristic of the desire to plunge into the origin, which was also found in Dubuffet. The artist sees himself as the demiurge ordering primitive chaos. "The fixing of a point in chaos constitutes a moment of cosmogenesis," said Paul Klee. This fixing "… confers on this point, which, in a primordial state, can only be grey, a creative quality spreading out from the center. From this, order thus awakened beams out in all dimensions." Unfortunately, chaos is simply dreamt by the painter. He does not dare to attack the real chaos of a traffic jam or inflation. And when Costa brings back to life prehistoric instruments, bones, prints, we have the same concern to rediscover the primitive, origins, and the untainted. In all cases, escape is needed from an unacceptable world. Likewise, where one says that the paintings of Tal-Coat spring from the abyss (Henri Maldiney), when writers want "to plunge into the abyss," (according to Gide's formula, let us not forget!), then we have the so-called plunge into original chaos, which, in reality, is a refusal to exist in real time. Perhaps it is the common denominator of all the contradictory tendencies in this type of art: whether it is a matter of the Message (which is always the denial of this world and the intent to change it by means of a political aesthetic); or whether it is a matter of abstract Formalization (which pretends to have nothing to do with Meaning (*sens*), because meaning reinvests this society!); or whether it is a matter of the great invocation of Being (the painting of Tal-Coat is a painting of the impossible in that it places us in the realm of being: it opens to the realm of Being!). It is all the same! Art escapes time. And because the escape is all too successful, only one thing is accomplished: the slavish repetition of this time.

The confusion between art and life leads to a confusion among the arts. The old classifications, simply stated, of painting, music, sculpture, and poetry no longer hold. The frontiers between the arts dissolve. Boundaries are erased. Musical techniques are inspired by pictorial techniques. An entire body of musical notation is inspired by graphic art. Specifically musical techniques, such as serial technique, serve as the guiding principle for the *Nouveau Roman*, where it compensates for the disappearance of the narrative voice. Painting, which has rejected perspective, wants to deploy its own presence in space. The painting of Bernard Saby is a case in point with its rhythms and its proximity to the music of Pierre Boulez. Sculpture has become a kind of architecture (beginning with Jean Dubuffet's walls). The sculptor Fritz Wotruba tends to produce architectural forms by banishing objective representation. Musical compositions borrow their titles from Paul Klee, which is telling. Adorno has specifically studied this osmosis of one art into another.[7] The same holds for the polytonalities of Iannis Xenakis in the exploitations of all audio-visual means to produce an entire symphony. Gérard Singer, in his *Passage de l'autre côté (1965),* creates a "total painting in space, without frame or play, without beginning or end," a reconstituted landscape abolishing the frontiers between painting, architecture, and sculpture. In the same vein, one finds Vassilakis Takis's *Sculptures musicales*. The sculptures sing like musical instruments at the same time as mobiles create a universe of movement. In reality, with electronics and programming that are attached to these "creations," one is forced into a world of architecture-sculpture. Takis "sculpts movement and trajectory, balls are suspended in a magnetic field," in an unfolding that is perpetually renewed. Movement is not imitated by the skill of the painter; movement itself exists. All this is possible only through the application of advanced technical procedures. The media are new extensions of our senses and, because these media establish new relations between themselves, new relations are established, as Marshall McLuhan would say, between our senses and our media extension, having influences in unforeseen ways. In other words, we have the great metaphysical claim: art rejoins life, which translates into, "Life has become a form of art," which also implies that, "All the arts have fused into one another," (the former boundaries and classifications being purely cultural and arbitrary). Only a vast array of possible novelties remains. But, this situation (which furthermore mimics the sensual confusion of certain drug experiences) cannot be maintained except by the most advanced technical procedures. I would say, further, that the existence of these procedures produces the idea of the removal of limits.[8] The most characteristic result of the development of technical means in all domains is the inevitable production of a self-generating progression and a rupture of limits in a transgression. The confusion among artistic genres is nothing other than a specific instance of the blind and ineluctable application of *technique*. What is more, we see the metaphysical pretense of joining together all of life, which is nothing more than affirming what happens when life is absorbed, assimilated, and expressed through and by *technique*. *Technique* is the prevailing absolute. The process of accumulation or interpenetration among the arts is not an invention of a generation of geniuses but is, rather, the mechanical result of *technique* transposed into the aesthetic domain by skillful mechanics. The rest is ideological justification.

We follow the same process when we consider the fact of the transgression of geographic boundaries. There is no longer American art, French art, or Scandinavian art. Art is determined throughout by technical procedures. Art has become universal. This was particularly in evidence in the 1970 World Fair in Osaka. In societies that have achieved a similar technological level, the arts tend toward a unity of style. "There are no fundamental differences of style besides the differences of individual temperament, between the cut-up sculpture of the Englishman King and the minimalist construction of the Japanese K. Yukara, between the enormous sculpture of a fountain—the water disk of G. Baker— and the waters on fire of Isamu Noguchi." And J. Michel, in his report for *Le Monde*, increases the examples of similarity. First, this new art springs from a system of rational and objective thought (but we would urge here that life itself has been so absorbed by *technique*), which is characterized by technical procedures aimed toward aesthetic ends. Next, this art pretends to be popular while passing "over people's heads" in producing a super work of art. Finally, this cultural ecumenic is based on wealth. By the means that it employs, this art is incomparably more costly than the medieval cathedrals, while being merely an expression of the technical system in one of its most salient features, namely in transgression. It seeks to interrogate and protest by being revolutionary, revealing the primary contradictions we will encounter. Abstract art that breaks conventions pretends to be revolutionary, as we show later. Art in its most technical form pretends to challenge society, which is nothing other than technical. The most costly art claims to be the most popular. The will to challenge is radical, according to Jean-Jacques Lebel. The generation of artists who show their works at the *Biennale* in Paris claim to be a generation who "denies accepted values," and who "rejects dogmas." They want to be radical by rejecting the city, *technique*, mass production, consumerism, and all that is artificial. They, then, advance on a romantic return to serendipitous wandering, to the so-called "natural" state. This is the philosophy of Hip. Of course, they will find raw materials—bits of pebble and coal, a clod of earth, and so forth, which are certainly not works of art because they are unmarketable and cannot be exhibited in museums. But, they are a protest against—against everything. And, one must also admit, from this, it follows they are protests against nothing. This art is a total challenge to society and technique that ends up subordinating itself to *technique*, to all forms of technique. It is the art that advocates Non-Art and that claims to deny the Whole, which is nothing more than following the technical process itself. In reality, in the process of total questioning there is a disappearance of specificity in art—leaving aside the triviality of impoverished art—together with a new kind of esoteric technical creation in the quest for a new reconciliation between art and technique, while art is absorbed by the techniques of the technical system. Thus, all art is trapped between the desire for revolutionary protests and the technicality of all of its operations, including those that expressed artistic tradition and unique, individual virtuosity. We are faced with a curious situation: how do we identify art among all the pretense? Non-Art, Anti-Art … but art nonetheless! Sculpture is rejected, although metal is cast, laminated, tubularized, hammered, and exhibited before spectators expected to identify what is known as sculpture now in its modern guise. One hears screeching, complex hiccupping, and growling sounds that are systematically linked together in a concert hall. The listener is expected to understand this as music in its modern incarnation. This link between prior knowledge and the possibility of something perceived as art cannot be broken, but this relation can be purely subjective. Someone claims to be a painter or a musician as an identity and is then recognized as such by a group. This mutual recognition is an intentional awareness that enables the differentiation of a piece of metal in a museum from that which the garage mechanic throws in the trash because it is broken or defective. But the claim of someone who uses the most modern techniques becomes particularly harmful when the message is revolutionary; they fight against an art, an aesthetic, and a society that dates from the nineteenth century. They are unaware that using the techniques of the technical system only entrench them more deeply in that system, transforming them into pillars of the current society and not in the one they imagine and fight against. Don Quixotes no longer exist; neither the folly nor the wisdom of Quixote informs. Instead, only a pretentiousness supported by blind ignorance prevails.

Modern art is contradictory in its parade of styles that march at an ever-accelerating pace, which follow the contradictions apparent in *technique*'s evolution and rhythms. Some have wanted to see two great ages. McLuhan posits the ages of the machine and the age of electricity, and it is true that the art of the machine age (1800-1930) has nothing to do with the art produced in the age of electrical dominance. On one hand, we notice an explosion with the division of classes, of knowledge, of functions; everything divides. On the other hand, we witness both implosion and contraction: we find a unified field of consciousness and a reintegration of separate functions. Urban space staged by the necessities of machines is no longer relevant to the world of the telephone, television, telegraphy, and electronic remote controls. Our electronic extensions transcend space and time. Art has followed this revolution, step by step. We see revolution but no continuity between the Cubism of a half century ago so typical of the machine age and the undulations of the cool media in a fusion of all the arts. Radovan Richta ends up with the same hiatus when he writes of the machine age and the advanced technical age. Moreover, we must understand that impressionism, expressionism, and the art of Cézanne are illustrative of the machine age, but we see a revolution in the making. The technical period is not a continuation of the industrial age but rather an about-face from all the tendencies of the industrial age—the exploitation and the division of labor, rigidity, one dimensional linearity, centralization, hierarchy—are turned inside out, as Richta demonstrates. We must situate the contradictions of art within this perspective. However, we naturally continue to judge Picasso or Cézanne as "moderns," which forces a rending and splitting of modern art that produces works that move against the above painters. The *Ready Made* expresses completely industrial tendencies, but the paintings of Bernard Saby and John Singer Sargent have nothing in common with them. During the initial impact of the industrial world and early application of *technique*, Pierre Daix has shown that painters were led toward realism and naturalism, which reveal industrial sensibilities. Currently, realism and naturalism have been completely abandoned for the benefit of uncertainty that conjures up electronic contingency! One cannot absolutely classify in one great movement the many subjectivist and self-generating tendencies. In place of an analytical vision of appearances, a multitude of syntheses have been substituted while artistic expression becomes more and more abstract! Art today seeks involuntarily to rediscover its symbolic function through and by means of abstraction. This process, however, is no longer accomplished through symbols, which were transformed with the ceaseless pursuit of new techniques within the technical system.[9] Thus understood, quarrels about modern and contemporary art, about the figurative and the abstract, are no longer relevant. When Delevoy holds forth on the modern mentality as a force that takes root and residence in the human's inner being, he exemplifies Dalí's harsh aphorism quoted above. In fact, modernity is the product of the technical situation, but, since this is neither clear nor evident, in a mechanical way. What remains are contradictory sedimentary deposits in the technical milieu, which are laid down by preceding epochs and which continue to suggest irreconcilable oppositions.

Nonetheless, this evolution does not explain everything. Other causes for contradictions are intrinsic to contemporary artistic production and do not derive from the succession of technical structures in modern history, and here we find a major bone of contention. For some, literature and art are not the playthings of the well-endowed and dominant class but are imbued with a content born of the involvement of a creative artist in the struggle with his age. Literature and art communicate with ideology because everything has become political.[10] The variations on this formula are endless, but they all say the same thing. They warn against disguised propaganda (for the benefit, of course, of an explicit propaganda although not declared as such); they call for involvement with the ideology of the masses, and they rail against economic constraint, and so on. In its totality, we see here art with a message, which, behind its façade of many Marxist explanations, amounts to little more than an art desperately aligned with a society devoid of signifying power, one of the effects of the technical system.

But, in contrast to committed art, we find a counter current: the technicization of society leads to a disengagement from all forms of message, even that of abstraction, and this absence of message leads to a veritable hypertrophy of technical formalism. (Moreover, in this current of thought, there are at least two possible positions: for some, art must express the ineffable; for others, it must exclusively create forms. In one case, one could say that abstract art, "neither, in its means or goals, evokes visible manifestations of the world." The inner man is, thus, freed to produce reality as he feels it. The artist reveals the concealed world within himself. But, in the other case, artistic creation becomes its own end. We need only concern ourselves with the production of a text, a color, or a musical score.) One no longer creates anything; rather, one creates a form that has not yet existed. That is all. Artists of the committed stripe will argue that other artists are anti-revolutionary and are running dogs of the bourgeois order. Those of the abstract stripe will condemn their committed brethren as retarded and retrograde and mired down in past delusions, because there is no longer the possibility of any message in the technical realm. Here we find the major schism in the art of our time, which, in all its expressions, is torn asunder.[11] There is no single style. The hundreds of styles, however, can all be grouped in one of these two camps, the study of which is the object of the following chapters. Furthermore, this is the crux of the debate between "rationalists" and "irrationalists" in art. For art with a message, content is expressible and therefore rational. And this perfectly conforms to the tendency according to which industrial society produces rational values and which must express itself in irrational works. The message itself, for example, to the degree that it is political, is rational even if the work appears surrealistic. We face here, appearances to the contrary, the continuation of functionalism: the idea that a work of art is perfect when it fulfills an aim. Functionalism, which once expressed fully the relation of art to the industrial world, has been surpassed, but its mantle has been donned by art with a message. On the other side, the irrationalists, for whom everything lacks sense and beauty, take up the cry of those who have been called organicists, i.e., those who think the work of art should shun the technifying process through an integration into the *organo* in the return to nature, etc. (Sigfried Giedion, Lewis Mumford, Pierre Combet-Descombes). We cannot examine these theses here, but let us simply note the curious reversal that has come about (perhaps the result of the technical transformation of the past thirty years). The rational functionalists, who were the best adapted to technical society, have ushered in a line of revolutionary artists (art with a purpose). The irrationalists, who preached a return to the living organism (hence, art without a goal but expressing directly what is lived), had as successors the "creators of form," and hence were most easily adapted to the neo-technical world. This major division conceals another similar division: on the one hand, we are told that art should be made for a public who, at this same time, should cease to be passive consumers in order to participate in artistic creation. On the other hand, one finds artistic creation becoming more hermetic, esoteric, and scientific. But, the two often overlap. The contradiction is not always between the two opposing schools but, as we will see, is sometimes between the so-called school of popular art, which may exhibit an hallucinatory denial of established convention and the school of abstract art, with its purely formal approach, which may align itself with public sentiment. Nevertheless, there is truly the creation of two cultures, more now than ever. The mass media, the supreme tool of communication, act on the total population. At the moment they radically isolate these two groups: "painters, musicians, modern poets are subjugated by the camera and the microphone, the film, and the tape recorder. Do the mass media consider their audiences as excluded from the inner sanctum of the creative process? Does the age of the mass media, through its techniques, isolate the creative artists with its machinery and remove them from the masses? A very large apparatus is involved, which isolates two groups of protagonists—at the one end, the creative artist at the device and at the other end, the viewer at another device. Contrary to opinion, there is no common ground between the camera and its audience." (Schaeffer). "The mass media compartmentalizes these two cultures stationed at opposite ends of the apparatus … Nowadays, we create simultaneously the most vulgar products along with the most esoteric. From vulgarity comes the new esotericism (Pop art) and from esotericism a new vulgarity (op art)." (Schaeffer). Modern art endlessly proclaims its service to the public. One need only grant it truth and an "unalienated" environment aimed at reinstating the need to find oneself in nature and to enable a questioning of the milieu in which it is found. But, in the same exposition (Paris *Biennale*), one discovers the coldest, the most abstruse and abstract of forms, a political discourse and hyper-sophisticated megamachine. Thus, we have an occasionally brutal call addressed to people who no longer invent and who are pushed once again to create with the hermetic techniques of an overblown technology. On the one hand, it has been asserted that, thanks to the paperback, the encounter with one's audience is established and that the writer, in order to take advantage of this encounter, must "possess a sufficient social depth," together with commitment and an humanistic outlook (Robert Escarpit). On one side, the radical questioning of language, of discourse, of the sender of the message, leads to a complete vitiation of all that might be "said." Expressionism can be placed at the confluence of the two currents. First, it has sought to communicate a crucial and even revolutionary vision through the shortcut of elided discourse and an expression directly through the medium of style without recourse to narration. But expressionism faces two directions at once: it has produced a violent, critical and oppositional display, a questioning of everything, of the viewer first and foremost. This is a protest against this world, especially the technical world. In the other direction form takes the upper hand precisely because it has become the message. Dissonance expresses the evil of the times, the spasms, the intensities of our situation; the intentional incoherencies of form testify to the realities of our lives. But the public does not march to the beat of these cacophonies, these spasms, these incoherencies, which it finds disagreeable. The more unrefined the public, the less it accepts. Expressionism should have been the most democratic of aesthetic movements and the most revolutionary, but, unfortunately, through its melding, in perfectly technical manner, of medium and message (the medium that is now the message), produces an art that can only be appreciated by aesthetes who are up to date on the latest research and understood by intellectuals who have, themselves, analyzed the misfortunes of the era. Expressionism has been the ultimate flowering of bourgeois art. It is a tautological discourse in which the speaker addresses himself about what he already knows. The audience, which they always talk about, is really the artist and his cohorts who consider themselves the only authentic representatives of the amorphous masses.

∗ ∗ ∗

Art with a message and formalist art are the hot and cold taps, we could say, with all the possible permutations of the two we have noted above. This dichotomy affords a clue. We have, in previous work, examined at length the difference between technical systems and technical societies.[12] Indeed, one must not superimpose or confuse the one with the other. In brief, the technical system is *technique* developing little by little into an independent, autonomous, and self-reproducing system. But this system does not exist "in-itself." It does not exist in the abstract realm of ideas. It is situated within and in relation to a certain economic, political, and cultural mode, that is to say, in relation to a society. Although the latter may be extraordinarily influenced by *technique*, it does not, thus, become a simple machine or its analog. I have never believed in the automated society or in the megamachine (Mumford) or in the self-regulating society (Pierre Naville). Society still remains a human institution composed of men who, however manipulated they may be, still experience problems, difficulties, hopes, and suffering. This human society is both technical and technological. In it the technical system develops with its own logic and biases. The latter is effected at all levels by *technique*, which develops by throwing out its pseudopodia in all directions. But, despite this, human society and the technical society are not the same. Will this be a permanent or temporary situation? Is it possible to say? In any case, the situation will generate conflicts and incoherencies, but it is precisely to this phenomenon that dichotomized art bears witness. On one hand, we have art totally integrated into the technical system, an exact reflection of that *technique* in its frozen perfection, in its meaninglessness, in its neutrality, its indifference to pleasure, to beauty to suffering, to the intolerable. On the other hand, we have an art that shouts out society's disorder in the grip of *technique*, an art which struggles in revolt against what it no longer knows and which questions everything simply because it is the thing to do. This art only expresses chaos under the impact of techniques that disintegrate and distort man's traditions and meanings. Such are, I believe, the two foundations of the two contradictory expressions of modern art and the reason why they are so much of a piece, so meaningful, so dramatic (but only *dramatic*, not dialectical!) in their contradictions.

* * *

This great schism between art with a message and esoteric formalism provokes endless other contradictions in the art of our time but also clarifies this art in the light of these contradictions. We shall now enumerate some of these contradictions that are secondary but nevertheless meaningful in the profound and radical dichotomy of all modern art. And, let us emphasize and recall that our study does not focus on an interpretation of art criticism but rather on the explication of an experienced reality. We do not consider a metaphysic of art but instead a question of what art currently signifies. On the one hand, modern art no longer tells a story. But, on the other hand, it claims to transmit a message or a decisive piece of information that is engaging but which can only be portrayed in an explosive, reproachful, and shocking form. Modern art, at the same time, pretends to "distance" the audience (but can only do so in discourse) or attack propaganda (which is found it its immediate surroundings). Art is caught inextricably in contradictions that are so easy to sort out. In the same way modern art challenges aestheticism, and I'm well aware that in the preceding pages the reader's patience or sympathy may have been strained when I wrote of "formalism." It goes without saying that there are no longer any formalists today (but then what shall we call the person who explicitly declares and proclaims that he does not want anything but a creator of forms?). Aestheticism, that is, art for art's sake, is a bourgeois position that sidesteps the critique of art. The old conflict between aestheticism and realism—to escape the real or to account for it—has been left far behind. The aestheticism one finds in the American novel and in other disengaged literature paints a vivid picture of a society given to futility, consumption, and to excess production, of a society where class conflict appears to have been banished. This aesthetic reflects a society conscious of itself as aesthetic, thanks to design,[13] cosmetics, hygiene, sports, and nutrition. Let us put this aside. Aggressive neo-formalism is the type of aestheticism that is self-justifying. The notion of art for art's sake is banned from the language but is replaced by anti-art for anti-art's sake. Within this realm there is no communication possible except through the application of learned and sophistical rules to interpret pseudo symbols, anagrams, oxymorons, Mobius bands, abysses, metaphors, anti-pastiches, chiasmi, and dapplings … All wonderfully destructive.

The same proponents of anti-art proclaim the exigency of untamed spontaneity and will produce, thanks to sophisticated techniques, forays into the extreme. The painter Hans Hartung is a case in point. His works are dynamic yet without content; there are active spaces, force fields, rhythms and no longer just simply forms. Bernard Dorival notes in the introduction to the works in the catalogue to the 1969 show that this form of painting is located between constructivism and an expressionism that seeks an expression in a refusal to represent the truth of an inner world. This is done by "creating the impression of an immediate improvisation while at the same time interjecting a perfection that will overwhelm us" in a nonchalance claiming to be spontaneous. As such it is much more connotative and denotative than found art, action painting, and the "wild gestures" of American painters, which are in every way as technical while pretending not to be. Hartung shows that spontaneity, to be noticed, must come from *technique* at its ultimate. Finally, despite the apparent contradiction, Pierre Boulez says the same thing when he challenges spontaneity and improvisation as the norm in music and when he shows that this approach leads to meaningless over-simplification, to mere "gallant music" (and he cites Terry Riley as an example) but what other alternatives are there? Reliance on *technique*?

Indeed not! We have subversion but an organized and calculated one that draws on the essence of *technique*. Unprocessed ideas no longer suffice. "In other words, we bathe in sound and create pop music, where musicians believe that they 'communicate' directly. But, says Boulez, pop music is more an outpouring than a form of communication. 'True' spontaneity can only be the guiding thread, a form of highly refined technique: one must have music by professionals both as instrumentalists and as composers." For Boulez, the partisans of pure spontaneity in art are like the leftists in a revolution: they are the most reactionary. As he says, "They play society's game and become nothing more than its safety valves." Subversion must consist of blowing up institutions, in this case the orchestra, " … so that it is no longer a traditional and reactionary body but rather a renovating force of contemporary life in all its facets." Later on we will examine how this attitude is also completely conservative. (But here the question will be conservative in relation to what!) For the moment, I only emphasize the contradiction between spontaneity and technicity supposedly joined together.

Another contradiction: the rejection of the object, while at the same time pursuing an art that only produces objects. But, we will take up the problem of the object in the following chapter. Suffice it here to acknowledge that contradiction.

Another contradiction: the society of spectacle is violently challenged (without even knowing or understanding Debord's line of thought). But art becomes more addicted to spectacle than ever, and J. Michel, in his excellent reporting on the Osaka fair, employs continually the term "spectacle," which is not surprising! In reality, the only problem is that of integrating the spectator: it is generally believed that the spectacle ends once the spectator has participated! This is tantamount to believing that propaganda has ceased to be propaganda once it has so completely incorporated its message that it becomes one with itself. In this process negation replaces true acts of consciousness. One simply claims that one is creating "Non-Art," "Non-Architecture," and Anti-Literature, Anti-Poetry to solve the problem of the rupture with traditional aesthetic activities. And so one calls it simply a break with old ideas without considering the relation of Art to technical society and, moreover, without challenging the role of the artist in this society. In other words, one changes aesthetic formulas but carefully preserves the most discredited of political, social, or so-called "revolutionary" interpretations. One imagines one's self a revolutionary in the theatre by killing the word in favor of physical expression. It becomes a game of expressing emotions through shock without the mediation of words, without logic or thought to produce meditations that increase in intensity to the degree they are not limited or circumscribed by "sense" and clear consciousness … They become pure sound, pure color and movement, a theatre that no longer speaks, a pure nothing. The Rorschach Test is a model of the genre. "Spectacle is reduced to gesture and murmur." "A body without a voice, a voice without a body." Silence and gesture. How can one not realize that with the suppression of language in favor of physical expression we replicate what is produced in the technical world where everything is movement? *Technique* is action. (And that is why we hear so much about communication …) *Technique* is the transmission of a certain number of effects. At the very moment when research on language reaches its apex, when the word is subjected to the most subtle types of scientific analysis, then the spoken word is summarily suppressed in favor of gesture—which is the way humans duplicate the mechanical, nothing more. We shall take up this question again when we deal with the suppression of sense or meaning because *technique* has no sense or meaning. Against this we can measure the depth of *technique*'s impact: those who most passionately search for the new have only one avenue, which is the transposition of the technical process into human life. Those who want to express the most profound misery and suffering in the search for freedom and love are moved to reject what is specifically human, that is, the spoken language in favor of behavior and action directly influenced by the machine.

However, the contradictions multiply. Adorno has summed up the situation especially well in his *Ästhetische Theorie* by showing that modern art is placed between two contradictory orientations without being able to choose either one. Every work of art that seeks to change is revolutionary, but it is, as such, immediately vitiated. If one protests complicity one is reduced to silence, which is another form of complicity. We have already encountered the double-edged impossibility. Indeed, art that is truly innovative would be disorder and perturbation. To the degree that society is totalizing, the perturbation must also be total. Art becomes simply chaos, as we have seen. Authentic art must undermine and then be challenged and rejected; it cannot be understood, at least by the sadomasochists who are the post bourgeois,[14] and, in this case, it no longer has any value as art any more than the mirrors and ornaments of "brothels." The continuity of art serves this society and its suppression would serve it even more, but of this I am not certain.[15]

Art is in permanent complicity with this society, which is totalitarian and absolute *because* it is technical. And, it is all the more in complicity to the degree that it mocks and to the degree it embraces the extreme. Mockery is the means of making acceptable what one would otherwise fundamentally challenge. The bourgeoisie is charmed by *Endgame* or *Godot*, because, by all evidence, what is presented does not affect them. Mockery allows the side stepping of the tragic dimension of calling reality into question. Thus, when art intends to expose everything with mockery, then it is reduced to a circus, a theater, or to, "… taking someone for a ride." Painting, for example, stops saying anything without knowing it; mockery and outrage—showing a pile of cigarette butts as a work of art—cease to shock. It is nothing more than the artist's wink, unconsciously directed toward the spectator, saying, "Don't take me seriously." Thus, modern art, which reaches the heights of tragedy, is trivialized by the very form it adopts. Here, Adorno's two extremes come together: an "uncompromising" art meets a "conciliatory" art. "Uncompromising music recognizes that, despite everything, society has a right to music, even if it is an inauthentic society, because society also reproduces its in-authenticity, and thus, by its survival, creates the objective elements of its own truth." Such is the true dilemma, the insoluble contradiction that Adorno perfectly brings to light, and which is only resolved by the absence of a serious element in art that has been pushed to the extreme limit of seriousness. Thus, the environment of the technical society reduces art to a more desperate situation than it has ever experienced at the very moment when the means of this society have multiplied, with the result that art is led to pursuing divergent chimeras. Francastel is right to emphasize that modern architecture produces, simultaneously, the "cellular style" and the "open style," with the elimination of walls. We are witnessing contradictory applications of comparable means.

Modern art is, therefore, not only divided into two great currents—popular and formalist, as it were—but, moreover, art is torn by multiple contradictions. Modern art is an explosion in all directions, an anthill of "whatever" or "anything goes." And here we must avoid two tempting interpretations: we could say, on the one hand, that this multiplicity is a sign of artistic freedom and individuality, even though there is no freedom with the fact that the artist is conditioned by the technical environment, by the primacy of *technique* over sense or meaning, and by the incoherence of these possibilities. Everything is possible, which is to say nothing has any value. Everything is possible, one thing as much as another, because nothing has meaning or reference. Procedures that must be constantly renewed are used up and thrown away as fast as the public becomes tired of them and when the surprise becomes stale … There is no freedom for the artist when one considers the incredible proliferation of means that conceal the absence of message in its banality, its conformity, and its clichégenic nature. The hodgepodge of discourse conceals its nothingness, even when there is a pretense of message. In reality, the artist floats along at the mercy of currents or waves; he is pushed in one direction and then in another simply by circumstances and means … Diversity, the swarming of the anthill, is merely the impossibility of finding a style and a sense of coherent meaning. Wriggling impotence should not be confused with freedom.

The other error would be to presume a dialectic at work. The contradictions, many of which we have mentioned earlier, are not dialectical! There is dialectic when two contradictions respond to each other—as in an affirmation and a negation—and when the outcome of the contradiction is included in history, i.e., when the tension between the two cannot be resolved except by the transformation of time flowing into history. Such a resolution would be the creation of a new state through time and the dialectical tension that holds the two adversaries together in an unforeseeable whole. There is no dialectic that can play between these contradictions, because they are nothing more than incoherence, cases of "whatever" or "anything goes" and not an affirmation or a negation inexorably leading to an affirmation of a wholly other kind. It is truly the "whatever" that corresponds to the "all is possible." A style is an affirmation, but there is none today. There is no modern or contemporary style. There are a hundred rockets going off in all directions with no other common element than the utilization of the most modern technical procedures. This art is incoherent because it is essentially a disparate assembly of masks pasted on a much more fundamental reality—a technical reality that is a rigorous system—and yet is internally incoherent and out of reach of art. Modern art has become an epiphenomenon of this system. It exists in the technical environment and constitutes itself in relation to techniques. Art has no unity, no reality, because it represents these accidental, haphazard, and superficial modes expressing technical reality. It does not exist on it own. It responds to certain utilities of the system and attaches itself to particular expressions of *technique*, and, thus, is drawn in prodigiously diverse directions as a consequence of the inverse roles it is called upon to play.

Notes

1 Robert Delevoy, *Dimensions du XXème siècle*, Skira, 1965.

2 Guy Debord, *Society of the Spectacle*, 1965.

3 Rookmaaker (*op.cit.*) has shown clearly that in the second half of the nineteenth century two great but contradictory inspirations inform aesthetic research: on one hand, is the domination of sensory perception, scientific positivism, and on the other hand, is the claim of human freedom in protest against positivism.

4 These two tendencies are explained in a remarkable statement by Paul Klee: "The more horrible this world, the more abstract art becomes, and the happier the world, the more art is realistic." (*Journal*, 1915)

5 *cf.* for example, Franck Jotterand, *Le nouveau théâtre américain*, 1970.

6 See on this subject the fine analyses of J.-J. Goux, *Les iconoclastes*, 1978, that explain the relation between perspective and the development of capitalism and individualism.

7 There are numerous examples of this confusion. For example, the book by M. Serres, *Esthétiques sur Carpaccio*, 1975), tries to translate painting by means of text (therefore this is not the usual type of artistic criticism). It is a "deciphering of space and arrangement," which is perhaps the *linear* order of phrase and discourse: "To meet the needs of pictorial translation, writing becomes musical in order to grasp the tangle of signs in painting; discourse becomes fugue and counterpoint …. The author uses a variety of registers … (P. Ciret). Hence, a book, which is all one could want; it is painting and music, but not rationally ordered language.

8 This is very explicit in minimalism and post-minimalism: First, architecture bereft, reduced to a steel framework, dressed in glass, and visible everywhere, the art of *less is more* as its inventor Mies van der Rohe says. Then, going on to sculpture, we see a simulacrum of architecture and which sometimes takes on proportions that could be considered either small architecture or giant sculpture. And after that painting expresses only the structures and coldness of steel, and of flat colors.

9 This is more important than the opposition of two stages, according to J. Leymarie in *L'Art*. Avant-garde art at first found itself in conflict with technical industry, but today, it is in conflict with a much more dangerous adversary: the culture industry and its law of conditioning. This is true but rather superficial and does not explain the profound tendency of the arts that have burst forth in our time.

10 Among the numerous works on this subject: J. Daniel, *Guerre et cinéma*, 1972; Cl. Prévost, *Littérature, politique, idéologie, 1973*; D.A. Siqueiros, *L'art de la révolution*, 1973; G. Sandier, *Théâtre et combat* 1970.

11 One has clearly seen the rift and the impossible connection between these two orientations during an international colloquium on poetry at Knokke in September, 1976. This colloquium brought together four hundred poets and was characterized by complete incomprehension between those for whom the poet should, above all, have a social and civic conscience and those for whom poetry should be a "free verbal exercise." The upshot between these two groups produced black and white proposals but certainly not poetry!

12 Jacques Ellul, *The Technological System*.

13 Design: it is not necessary to read the book of Hoffenberg and Lapidus on design (*La société du design,* 1977), which is a redundant discourse demonstrating that design (as a kind of whatever) is a vehicle for capitalist ideology and becomes, through hyperbole, the instrument for setting up a code of value and a means for investing the object with the norms and social models of capitalist production. Enough said …

14 Jacques Ellul, *Métamorphose du bourgeois*.

15 I believe that Jimenez's book, *Adorno: art, idéologie, et théorie de l'art* has not truly taken account of Adorno's dialectical mode of thought. The author clearly has not understood the basic problem when he tries to resolve it by reducing Adorno's work to a "paradox of art," lifting it to the universal, and making it a complaint about the takeover of art, and inviting one to see his work as an act of refusal. All this is very superficial.

Chapter 2
Art in the Technical System

By placing art in a very direct relation with technique, by situating it in a technical milieu, by making it depend strictly on a technical system, it is not a question for us to reproduce a kind of determinism of art in the mode of Hippolyte Taine's theory. There is a two-fold difference between his kind of explanation and what I here propose. First of all, I am definitely not offering a permanent and universal definition of art. I am not generalizing from what we can observe today as a concrete universal. I do not pretend to deal with the process of artistic creation in it itself nor with the eternal function of art. But, on the contrary, what I observe seems new and different. Our situation in relation to the technological world has no common measure with what has preceded it. One needs only to be aware of the fact that, up until now, all the explanations, interpretations, and significations advanced on the subject of art are devalued and passé. The other difference is that I nowhere posit a direct conditioning of the artist by circumstances, the milieu, the environment, etc. to make what he produces. I nowhere deny the autonomy of the individual, but I try to explain the actual situation of the artist and artistic creation in this environment that is new. And so it is less a question of direct conditioning than of confusion, resulting from a disorientation and uncertainty deriving precisely from the multiplicity of functions that art assumes in a technological society.

The second point I would like make is this: I will not demonstrate what determines modern art but rather explain why it is what it is. That is to say, I will investigate this new so-called artistic freedom to see how it plays out through various codes, schemes, and media. Hence, I try to clarify, finally, what modern art accomplishes, what it reveals in our present world. Thus, I do not try to explain modern art in itself but rather its meaning. It is not question here of absolute determinism but a question of how the ensemble of means, correctly understood, is revealed precisely in the play of different creators. And their differences are definitely more significant than their common origins. It is the unconscious differences, not expressing freedom, that best describe the world where these differences flower.

I / The Relation to the Technical Milieu

Let us begin with the simple idea of art as the mirror of the environment. Let us not ask what place is left for art (it is a false question) but ask what art expresses and in what measure it reflects this society. Has the relationship of a work of art to its environment been essentially modified by the fact that now that environment is technical? Francastel can say with total correctness, that, "Modern art is not based on a capricious play of provocative forms; it does not develop far from experience, but is based on the total activity of contemporary humanity … Artists do not play a social role of isolated individuals independent of technicians. Instead of the concept of separate histories of different disciplines and human activities, one needs to substitute a total examination of the types of expression in a society that models itself by expressing itself." The whole, if we retain the idea of a mirror or of expression, leads us to ask: a reflection *of what*? It is essential to note that naturalism and realism are born at the moment of industrial development. Émile Zola codifies an art copied from external reality, an art that seeks to be scientific and neutral. But, up to this point, art has always been a reflection of the natural world. Man was bathed in the natural. He only lived with air, water, rain, and trees. Even when he lived in cities. Even in his strangest creation, in his most unbridled aesthetic, he always took these elements as a point of departure. And man, taken as a theme of the artist's representation, was of nature, submerged as a participant in nature. Now, one notes that, at the moment of realism's inception, painters had a tendency to separate themselves from the "laws of nature." They stopped treating the human as a natural reality. Claude Monet, or Claude Pissarro, or, even more, Paul Cézanne, took nature more as a *theme,* but in such a way that their themes were neither reflection nor mirror. Their aim was not to express this nature, or even, according to realism, to understand it better; their aim was to paint. That was all. This implies not only a rupture with the milieu but that the milieu had changed. Nature still was held as a common point of departure but was no longer the milieu of painting, and that's why it could take nature as a *pretext* for plays of light, and form, etc. In reality, the milieu that gradually came into being was at first a universe of the machine and then an urban universe. But, this is still completely superficial. It was technique itself that became a milieu, in the strongest sense of the word. That is to say, the place where one finds the possibilities to live, the orientations of one's life, that which surrounds us totally, and that which we are obliged to know before a knowledge of what is the other. Technique surrounds us like a total and encompassing cocoon, which renders nature perfectly useless to our immediate judgment, a nature that is dominated, secondary, and meaningless. What is meaningful is technique. Modern art is still a reflection but of a reality that has changed, and if one portrays man in a novel or in a painting, it must be of a man who is no longer situated in the natural milieu or environment but in the technical milieu. One understands nothing if one insists that there is no other source of form than the visual expression of the natural world. (Sweeny). Nature has been torn apart and disassembled by the sciences and technique, but that's not all. Technique has constituted the total milieu where man lives, moves, and has his being. All his strong impressions come from technique. Art is no longer in any way able to "represent" nature; and neither can the artist and nor can modern man.[1] One other error would be to consider that, from this break with nature, art no longer represents anything but itself, that it dives into the subconscious, or, yet, that it is only the rootless product of canvases and texts. In reality, this is the illusion that comes from not weaning oneself from the natural model. "Since painting no longer reproduces nature, it solely reproduces painting 'in itself.'" Or so they say. Nothing could be farther from the truth. This art simply expresses through all its tendencies, its interpretations, and its visions, the technical milieu. This of course does not imply that the poet speaks about the machine or the painter paints machines.[2] But that everything is fundamentally inspired by this milieu of which modern art is the translation.

Further, one must not mistake the subject of this milieu. One cannot simply state that the problem is a "social consequence of technical progress or of the dehumanization of the modern world." This is still linked to a naturalistic interpretation of Technique, where one applies criteria derived from "nature" in order to judge art. (It is not normal for a face to have two eyes on the same side as the nose, etc.) Art continues to constitute meaning and value but meaning and value of something else. It is still the metaphor of a structure where the relations are always more important than the things placed in relation. Nevertheless, the things placed in relation are not unimportant. These are the things that define this art. And the manner of the relationship, the working out of the structure, depends absolutely on the milieu. Under these conditions, if one still holds as a pre-established concept, as a cliché, the relation to nature, we would have the impression that there is an essential challenge to reality by modern art. One will still claim that Cubism is a process of transforming natural forms, but what kind of process, and why? And one will compare it to Surrealism where forms no longer intervene, where it comes to an alteration of meaning, an unveiling of an invisible universe. But what is the meaning? Abstract art, they tell us, contests the reality before which man is placed. We learn that our relation to the world is many faceted. The new forms created by artists are the expressions of a new experience of the universe. And one reduces everything to the portrayal of the mobility of objects in light, for example. Now all of this still supposes that one lives in the milieu of nature, or simply that we have learned to see things differently. One will insist on the processes, on the creative imagination of the artist. One will say: "There is no longer an adequate way of portraying a *given reality*." But this is because reality is still nature. "Modern sensibility frees itself from reality." But this reality has changed. The act of referring occurs in relation to the milieu. The real is not what is given to us in the "universe," nor, even more, what we construct abstractly, but rather the artificial technical milieu. The real, in relation to what art seeks to convey, is the real constructed by our own means and methods. It is identical in function to the former natural milieu. There is certainly not, as Delevoy says, a "defiguring of the real" or a dematerializing of the real. One tries, once again, to bring back to nature new modes of observation and reflection, but, they refer to something else. The radical change in modern art is not a product of the artist who would be an original creator, an inventor of a new world. Rather, this change comes from the work of the technical milieu, which carries with it changes of perception as well as of pattern, which places art in contact with a reality qualitatively different in its structures, in its processes, in its forms, in its space and time, its levels of appearance, and its rationality, etc. And the artist is only the faithful transmitter of this transformation. Nevertheless, with all the disorder that distracts him from continuing to deal objects and themes from nature—trying to maintain a naturalistic mentality—only rarely does he dare to engage this new universe; he is like all of us, betwixt and between, and continues to eye distractedly the fields and the rivers, the sky and the ocean. But he no longer sees them except as remote images—almost mythical, of a nearby reality, pregnant— that sweep him up in a current that is anything but a rushing, mountainous stream. This art seems less the bearer of truth than the herald of distortion with the belief that the artist continues to maintain the relation to nature that was held in traditional society. And these sensibilities are completely out of place in relation to this new milieu. The artist continues to dig into his stock images and symbols, but nothing corresponds to them, and nothing hits the mark. He is torn between traditional sensibility and the technical world, which in relation to the former is nonsense. And the rupture is so much the greater, when, as Francastel has well noted, Science, like Art, is still *figurative*, while technique is not. The technical universe places man in a non-figurative world of processes and means, but a world still made of objects. The objects are simply silent witnesses. But if technique is above all means, it is wrong to attempt to situate it in relation to art in the following manner: wrong to say that techniques give the artist new means of self expression. But technique is not a means. It is an ensemble of means coordinated with each other and constituting a world of forces in a new world that completely replaces the old, a milieu of means instead of a milieu of living organisms. The objects, produced by technique, are not linked to one another organically or systematically; the technical universe is essentially a correlation of means. In the natural milieu, functions were in balance, whereas in the technical milieu the growth of means is unbalanced and explosive. We manage to construct a unifying sense that is totally devalued (hence another factor in the break-up of the arts). "Art can no longer signify society because there is no longer a transcendent and unifying principle for this society." (Jean Leymarie). But it signifies another milieu, and if this milieu appears meaningless, it is because we do not yet understand it. We are not yet accustomed to the signals and shocks of this milieu of means. We are unable to locate our communication outside the natural milieu, hence our difficulty in symbolizing.

In contrast, symbolization in traditional societies was man's major mode of apprehending and influencing his milieu beyond a simple material mode. The function of symbolization had become one of the major modes of action. And what we call art was one of the essential forms of symbolization, as, for example in religion. And now we find ourselves placed in a milieu resistant to all symbolization. Technique cannot be symbolized for three principal reasons.[3] First, it has become the universal mediator, and because it is itself a means—in its capacity as means—it is not the object of symbolization, but rather it is also, by its power, outside of all other systems of mediation or symbolization. It is, in the second place, a producer of a communal sense. The communal act today no longer relies on the support of the symbolic but rather on a technical support (the play of media, for example). Simply, technique establishes a non-mediated—an immediate—relation with man, who, in the past felt a strong need to distance himself from nature but technique seems not to require such a distance. It seems to be the direct extension of the body. Who has not heard it said that the tool is merely an extension of the hand? Thus, we pass from an *organic* world, where symbolization was an adequate and coherent function in relation to the milieu, to a technical system where the creation of symbols has neither place nor sense. What symbols are necessary are produced out of technique itself. Television or advertising offer abundant symbols of technique but these come from the very working of technique itself. Therefore, the technical milieu is never understood because symbolization is excluded. And, from this fact, art, the foremost minion of symbolization, finds itself chaotic and torn between its "vocation" and that to which it can no longer aspire; an environment made up of discrete pieces belongs to structuralism but not to symbolization.

The other major difficulty rests on what one could call the anomie of the technical system. But I do not hold that there is incoherence or chaos, but rather that the system obeys its own law of development and organization. The anomie springs from the clash and incompatibility between its autonomy and the moral law of the individual, on the one hand, and, on the other, society's traditional systems of signification. Characteristic of this anomie, in the domain of love or sexual relations, we find a tendency to snatch them away from the social in order to establish them on a purely aesthetic plane. Amorous or sexual aestheticism is absolutely characteristic of the technical age.[4] And this holds true in all domains: we will see later on that this anomie produces formalism and aestheticism, which art must engage or pretend to disengage. One can, of course, conveniently contrast, classify by styles, and attempt to "rediscover one's self," in cubism and the non-figurative, for example. But, in reality, there is a common element, which is the relation between art and the technical milieu. The non-figurative betokens a relation to nature—to what men see and hear—but he no longer hears singing birds, or the sounds of cicadas, or thunder, or the crashing of waves. He hears honking and the rumbling of engines. Only in art does man still conceive reality through ancestral stereotypes. The "non-figurative" is in reality the figurative manifestation of the technical milieu. And even art with a message, ideological art, is itself forced down the road of abstraction and non-figurative representation. Art no longer proceeds by direct communication in rational and well-ordered language. It is forced to produce emotions and commitments by short-circuiting discourse and by means of signals and shocks, the charts and graphs, or the visual clichés of the technical society. When modern painting moves away from the figurative, it approaches the same kind of rupture as the theatre of the absurd or atonal music, which is music tied to the new milieu.[5] One can truly say that abstract art is, "stretched by a dialectic between itself and the alienation produced by historic conditions," (Leymarie) but this is only explained by the extreme difficulty of this art to reestablish itself in a radically strange milieu. The relationship, philosophically so often affirmed between Art and Technique, no longer signifies anything from the moment one has left the era of the technical operation to enter the order of the technical phenomenon and then the technical system. Hence, one can understand to what degree the usual appreciation of abstract and of non-figurative art is a mistake; this mistake is upheld, furthermore, by the artists themselves who do not know what they obey, giving us an outstanding example of what Marx has analyzed as false consciousness. This habitual judgment is that "informal aesthetics is grounded in accidental and subjective values; the material, the gesture, the written word, the random motion, all that reveals the traces of the hand … a field that is fluid, infinite, and gives to the work its substance and meaning. Informal art is the first complex style of which no element is determined in advance, where the sign precedes what is signified and postulates its ambiguity." (Leymarie). Of course this is also formally correct but essentially this art is not all random because it is directly conditioned by the technical milieu. The work of Robert Rauschenberg, the father of Pop art, which we have seen exhibited in Paris in 1968, is completely relevant in this regard. A work that mirrors society, thanks to the discarded products thrown away by this society, a work "questioning everything," as in an exposition of white fabrics alone without anything else, is a perfect example of nonsense. An "opposition" to consumer society that transforms itself, little by little, into works of a technological nature by introducing "reality" by an image taken from current events—a helicopter, a photo of Kennedy, and so forth—silk screen images on glass that appear and disappear, automatic devices used for effect; the valuation of objects devoid of sense is the sense of nonsense. The pictures printed every day in newspapers are, in Rauschenberg's view, the most interesting images. Hence, we have an insertion into the technical milieu, the utilization of its products, a vision of the "world" as a vast continuum, which in fact is borrowed from the technical system! It tells us, in effect, that *the city* is a milieu without beginning or end. Painting has become a gigantic hallucination of this milieu, a non-composed field of images, an uninterrupted flow, with no beginning or end … the spectator must be "inside it" … as in the street. And everything must continually change, especially in medium, which permits recreating the very language of the work (a position specifically held by technique!). The true problems are those that spring from the manipulation of objects and of the physical order! But Pop art places the informal scholar on the terrain of technique when informal art attempts a "global structure derived from the mathematics of Georg Cantor and Michel Loève." Here, we no longer have throwaway culture. By contrast, we have the challenge of experiment for experiment's sake, which is the battle cry of Pop art. But this is no more than a surrender to one of the currents of the technical milieu when it unites in a single formula the "surpassing of all that is possible" with "a dyed-in-the-wool generic pantheism." In a nutshell this is technique's aim! And hence, we have neo-realists feeding off technique: for example, the influence of advertising, electric lighting and luminous signs to produce a futurist spectacle "the noise of the technological great beyond," the organization of a surface produced by an electronic machine, the utilization of magnetic force fields, etc., and here Vasarely offers us a masterful lesson in how the technical milieu operates: all his renderings are based on a combination of *ordre et corpuscule* (wave and particle) hence the aim in a very technical way of continually bombarding the spectator with undulations, segmented lines, networks, swerving surfaces, expanding and contracting spaces, geometry and the laws of science, interpreted by the artist. The effective reality of our time is effectively assumed and projected by the artist in a truly new figurative realm. Thus, the artist translates, without knowing it, the milieu in which he lives. I clearly say *without knowing it,* because he truly believes that he is producing something else. He believes he is calling into question this society; he believes he is inspired by science; he believes that he is the demiurge of motors, lights, and magnetic fields; he believes that the basic problems are political, and so forth, but he is totally unaware of the decisive factor: *the technical milieu* as a system. He continues, instead, the unconscious and mystifying work that he claims to deny. And by the same token, of course, the general public, which is no more conscious of the technical milieu, gives this art the cold shoulder (including the so-called popular art). Music, like painting, has come into conflict with habitual needs. But, here, pop music has enjoyed an extraordinary success to the degree that it provides an apparent escape route from this intolerable world, as a road into the irrational and the dream world and as the great rejection. But this was an illusion. Serious music, which, by contrast, tended to reveal the real, collided with the customary ideas of music held by the general public, and it became annoying to the degree it created awareness of what the public would want to ignore. This aristocratic music (Arnold Schoenberg for example) is typical of the impact of technique: "It sketches the image of a state of affairs, which, for good or ill, disregards history." Atonal arithmetic music is just as rigorous as any technical product and challenges appearances, superfluities, and ornamentations; nevertheless, the attitude of its practitioners (as expressed by Adorno) borders on superstition, magic and astrology. The apparent opposition to the rationality of technique produces in the practitioner an irrational type of religious behavior.[6] Enough said.

Now, everywhere people yield to the obsession of no longer representing things directly, because the direct representation of a machine in motion was good for the machine age, but this is no longer the case; what constitutes the technical universe is less and less directly *representable* because it comes to us directly without mediation. And here we find the legacy from 1920s' Russian Constructivism; art no longer reproduces but constructs; the traditional painting created the pictorial form of the object given by the external world; new painting creates the object itself (Nikolai Tarabukin, *Le dernier tableau,* 1972). The aesthetic mission no longer has an ontological value but only a semantic one that conveys the permanent exigency of change and progress. But this can only be carried out in the production of a thorough-going facticity. Adorno is correct when he writes that "the artist's mastery of nature should itself appear as nature," as for example in Wagner, who combined "the underlying irrationalism with the rationalism of a conscious mastery of means," (typical of the technical system). This inevitably leads, given the world where it develops, into a type of music in which "poorly done" replaces the old concern for the "well done." "His calculated errors become apparent in the open contours of many contemporary paintings … which disassemble or deconstruct all unity of pictorial configuration. Parody, the fundamental form of music aping music, aims to imitate something and to satirize it in the process." One can only satirize what belongs to the natural order, to the world in which we used to live, because this world—including that of the so-called natural sentiments—is finished, liquidated by the technical invasion, and all that is evoked is at once outdated and painful. Art can no longer seriously express anything but technique, and the artist can only be a technician (but, once again, I don't mean by that that he should only possess or utilize the specific techniques of painting or of music; it is the question of the technical mentality [*la mentalité technicienne*] and of a reflection of that world. Consequently, he can only flout nature and produce and reproduce facticity. And this often carries the impression of a tragic "fate," which is indeed an apt representation of what our universe has become. Andy Warhol partially demonstrates the "American way of death" with a flood of technical signs, soundly denouncing this world, with probably a pang of conscience, but at the same time with a very characteristic cynicism in the pursuit of success at any price. Thus, the very being of the artist is the echo chamber and the product of the technical world. Hence, he furnishes, through himself, a reflection, an echo, a radar blip of that reality that we refuse to face. The dissonances of music speak to us of our own condition in this milieu. We take in, with Schoenberg, by the intermediary of music, the movements of the unconscious, real and undisguised, grappling with this new intolerable milieu. We see, we hear these traumas inflicted by the technical universe. The assault on our thought as on our sensibility, by mechanical processes, is the decisive fact. On the one hand, this leads to a kind of art that one can call mechanical, capable of mastering the combinations or exploring the possibilities by a basic originality (Moles), and, on the other hand, to an art of confrontation, of deterioration, which expresses the unhappiness of man hypnotized and subjugated by the technique he believes he has mastered. The rending of art finds its meaning and at the same time modern art finds its unity in this reference.[7]

But, one should mention a completely different interpretation. Jean-Joseph Goux,[8] who has made a study of current iconoclastic movements, starts with the affirmations of abstract painters who claim that their work is a spiritualizing process. On the one hand, he seems to accept this act of liberation, and, on the other hand, analyzes it as the expression of the passage from an individualistic bourgeois society to a social collective based on "operations." By breaking with nature there is an iconoclastic destruction of idols and a spiritual liberation. This is accompanied, for example, from the movement of painting to sound through the language of forms and colors and a direct access to the soul. This is a great spiritual transformation announcing the advent of the reign of the spirit. And Goux relates this de-figuration with the axiomatic tendencies of mathematics and Saussurian linguistics, "which makes of language a system of pure values without any root in things … and with the discovery of abstract spaces. … All this would be nothing more than putting a certain way of painting in relation to scientific trends, if there were not a certain tone of triumphalism in a march toward mysticism: 'Art, which the soul uses in order to act upon other souls.' 'Art no longer passively reflects the real but affirms the inner constructive and organizing powers of the spirit.'" "Abstract art marks the outer limit, the idealistic apogee of detachment from nature and from the material world … Art essentially prolongs the organizing powers that we still call 'spirit.'" And Goux shows clearly the spiritualism of Kandinsky, the theosophy of Mondrian, the "Nothing Hidden" of Kazimir Malevich. Of course, I do not hold that Goux adopts this point of view; he attempts to explain that the movement of dematerialization must not be interpreted from a spiritualist horizon. This reversal transforms the sensualist conception of matter and nature, and puts it in line with Marx where, according to him, the notion of matter corresponds to sensible and practical activity. Abstract painting, like modern music, is the organizing principle of living forces. And the passage from an external perception to an internal reflection allows one to reach the source of the senses or "the effect that the source has on the production of the senses." All of this is most interesting, but Goux does not find an instance where representation has yielded its place to operation; painting is in no way a result of the painter's freedom but is the substitution of the technical for the symbolic. Without doubt he makes reference to technique but in a very superficial way. Technique makes the production of the new possible and, furthermore, "scientific and mechanistic activity as a driving force of transformation and production, constitutes the new blueprint that abstract art adopts and that is, in effect, a significant player." I completely agree with that, but from this he draws no conclusion. And he is unaware of the reality of technological society. Technique remains for him a socializing factor, and abstract painting is a correlative to a social relationship where the creative elements are set in motion by the interrelation of social practices: "A social relation extensively dominated by the transformative action of organizations and apparatuses and no longer by individual-centered decisions … is a socio-symbolic system where the concern for the origins and dissemination of things dominates the idea of a thoughtful creator. Rationality is no longer mere marketing but is epistemological and technological; rationality no longer seeks equivalence, value, reflection, reproduction, essential being, but operates on what is anterior to these things …" Here I share my surprise and disagreement, because he is not describing the triumphal march toward freedom through iconoclasm but reveals the radical blockage of processes outside of man over which man no longer has mastery and that only produce what is incommunicable, what is nonsense, and what manifests his own absence. Man is replaced by the technical process, which operates on the most profound level.

But, here, in these conditions, art, so heavily influenced by technique attempts to become, in its turn, a complete milieu, a structure, a counter-milieu, an environment. In other words, it attempts to reproduce for its own purposes what technique has effected in reality by refusing explicit reference to a milieu. Music tries to be an artificial construction of a truly non-figurative sound environment, produced in waves by machines for infinite consumption. The sculpture of Louise Nevelson aims at a complete environment. Her work is not to be examined but lived! The sculptures are walls whose multiplicity and clever incoherence exceed the limits of sensation and description—a structured universe no longer symbolic but traumatizing. Art is not a décor but a milieu. An astonishing pretense because, unfortunately, this milieu exists in a well-defined space, as part of an exposition, an unpacking of works, which the viewer, willingly or not, must pay to see. And so we are far from the creation of a substitute environment. Without doubt, architecture is better situated in this domain, and one could test this neo-milieu with the urban machine of Beaubourg. It is a machine for communication where ideas are to be exchanged between a creator, a public, and as Jean Michel so perceptively says (*Le Monde,* February, 1974), an army of critics. The Beaubourg machine is meant "to assume the function of the museum-cultural center, to receive and process information, and to vomit it out. It is not a question of a museum but of a kind of tympanum that retranslates a culture that is given data daily. The center, a communication machine, will also be a machine for circulation on an urban and mechanical model: open stages, places, forms, and an architectural machine converted into a spectacle …"[9] But we run up against the same obstacle: a limited space where visitors, spectators, and the curious congregate. The hoards who stroll through the Louvre or Beaubourg do not change, and art continues its true role as confidence man, which is all the more serious because it presents itself as true. The terrible illusion consists of thinking that art can provide a counter-culture to the technical world and produce a counter-milieu. We know that this is McLuhan's position. But there are two sides to this: either it is a counter-milieu as an antidote, forming the perception and the judgment that it provides a critique of the technical milieu, but this seems impossible if this counter-milieu, as McLuhan holds, is constituted by television, is produced by technique, and indeed transmits the very power of techniques. This confirms the insertion of man into the process of technicization. Or else, as McLuhan also holds, art makes us conscious of the psychic and social consequences of technique, and this corresponds to what we said earlier, but this is not really a counter-milieu, for this assumes that, deep down, man, indeed all men have a permanent essence. From a certain point of view, television really constitutes a counter-milieu, but it is not the only influence on modern art. It is only one aspect of the media influencing the arts. Indeed, television affects our vision of the world, which all arts seek to do, but it is only to confirm the vision that the technical world gives us.

To conclude, we therefore come to this concept of an art that directly expresses the technical milieu. Its diversities, its contradictions, express, on the one hand, the extreme difficulty of a radical transformation, for it is not easy to pass from one milieu to another. It expresses the difficulty for each and every man. On the other hand, the extreme diversity in aspects of this technical milieu plays violently and incoherently with its powers and means. Finally, it is impossible to follow the rapid variations of this milieu. Basically, technique is an Art that expresses and has its foundation in science. This sort of conviction that modern art is an "incarnation of our scientific credo," that it is the scientific discovery of matter and light that is translated into modern art, that scientific research delves behind appearances, is what science has taught us: that all this is a simple *a posteriori* construct of intellectuals. One may notice that painters and musicians have worked in this vein, seeking correspondence to such and such a scientific theory. But, this is the kind of reconstruction that depends on simulated correspondences. The true rooting of modern art is carried out in this new milieu, which is both real and determined. And the passage from a traditional, ancient milieu to the technical milieu is sufficient to explain all the distinctive traits of modern art.

II / The Object

We live in a universe of objects produced by technique, a mass, a mob of products, which, through the power of technique, resembles an invasion. And these objects are all the products of Technique, one of the essential characteristics of the world in which we live. We have already said that design attempts to capture, from an aesthetic point of view, this world of objects, but it is a bit of a dream to believe that artists could lay down the law on the technical milieu. This law only holds for a certain type of object in our environment, i.e., objects of luxury like the Dupont cigarette lighter. "The art of living and inhabiting" is not an art for the masses. Here we speak of aristocratic art, and the mass-produced objects of industrial production, say of plastic, seem far removed from any aesthetic image. But, as J. Michel has noted, boundaries are imprecise "between these cultural objects," which are hung up in galleries, turned by motors, dazzling with crazy electric lights …, and these "commonplace objects" like easy chairs, tables, lampstands, drawn from the same aesthetic sources, all of which appear to have a coherence of style. Currently, it is impossible to situate oneself against a framework of luxury objects, on the one hand, from a framework of utilitarian objects produced by technical means, on the other,[10] a completely fallacious distinction. In reality the proliferation of objects has led away from the distinction between the everyday and the artistic. Marcel Mauss held that certain objects were specifically produced with an aesthetic end in mind; other objects had a different purpose—hunting, housekeeping, magic, and so forth—but had an aesthetic value as well. This is the classic view, which is now upended. There was Francastel's objection that one cannot define beauty as a power beyond the object, isolatable and additional. (Let us remember that Francastel still clings to beauty as a criterion of art.) It is impossible to add the artistic quality to an object, and it is impossible that an object be only an aesthetic expression. The object is the production of the total man. There is no scale of absolute beauty in itself. Rather, art has a double speculative and operational character that cannot be divided. The object of art is a complex whole, which excludes a hypothesis of absolute specificity of what is art or non-art. One cannot separate the aesthetic function from the general techniques of production or from all the surrounding utilitarian or representational objects. One must also reject the idea of an immanent beauty in the rational production of utilitarian objects. There has been no contradiction between the creative developments of mechanism and of living art to recall the view of functionalism that was characteristic of an epoch or age when technical production was welcomed with enthusiasm. Then, technique was considered liberating and forward looking, and the universe of objects produced by technique could not be separated from the universe of objects of art. One had to acknowledge the relation and reintegration of the one into the other. But suddenly man retreated in disgust at this invasion of the object. He saw and felt himself reified by proliferating objects with a violent reaction against consumerism. Then, the object of art became, in turn, a point of challenge and argument. It was no longer a question of art penetrating the desired object nor of affirming the coherence between the two: it was, rather, a struggle against the deluge of plastic products and consumer society. The object became an obstacle for a new interpretation of past schools, of impressionism, of cubism, of abstract art, that offered a negation, a challenge, to the object. "For centuries the object of art resembled a bull in the arena." (Roger Bordier: *L'objet contre l'art. Toute l'histoire de l'art moderne est celle des vicissitudes de l'objet,* 1972). Art has tried to conquer the object and even to eliminate it. Lately one sees products which melt, which self destruct, etc., to ensure the activity of the artist would not be recovered. Artists "intervene." They propose "objects of reflection" like a container of "pure" air in order to make one "reflect" on pollution. All is questioned, including the very production of what is presented. The critique of the object and the refusal to produce objects coincides exactly with the complete subordination of modern art to technical mentality. Again and again, in recent times, people are obsessed by the invasion of objects; people have made a system of objects: but in reality, it is, simultaneously, the complete devaluation of the object to the benefit of the means of action and to the benefit of techniques. Hence, this general observation (and for further development I recommend *The Technological System*) is confirmed by the tendency of art to become, once again, an example of technical reality. Modern art does not seek the finished product but rather the validation of the means producing that object; it erases the object in favor of the method. One does not write a novel but a text on the act of writing a novel, nor does one paint but rather one illuminates the technique of painting, and cogwheels themselves become works of art, in what would have formerly been called rough sketches. Therefore, this not only expresses the primacy of means and the obsession with means, which occupy the entire mental field of the musician or of the architect, but, furthermore, the idea that the product has no importance. In reality, Technique allows *everything* to be done, and since all is permitted, the objects of aesthetics become all or anything, a totality of whatever.

It is not easy to eliminate the object from Art! And furthermore, I believe that Bordier has truly seized on one of these dilemmas: the challenge of the object but the impossibility of doing anything without it! "The object wins: it can be at the same time its specter and its body, its reality and its metaphor … it remains what it is …" One can neglect the object as a model, but it returns surreptitiously against the will of the artist (a spot can always represent something!). Painting is brought back to the reality of objects; the objects of music are sounds. One can take the object completely off the canvas and it affirms itself with renewed force in the mind of a spectator preoccupied by its absence. The artist always ends by capitulating. "It is the revenge of things." The object then presents itself as artistic in its own right, animated and imposed in its raw presence. A gallery calls itself "the raw object." It will be art itself: Pop art. Pop art is the capitulation in the face of the invasion of technical objects. The work of art becomes an advertising poster, the recovered everyday object, a simulacrum of a landscape made of discarded objects, a computer screen. There is a fetishism consisting of integrating into art visual and auditory elements of our environment. Everything that makes up everyday mythology is transformed into art. Pop art has managed to propagate its work in the same way as industrial products, and so we have seen a subculture that refuses classical culture, perhaps, but this is done in order to assimilate itself within the production of technical objects. Here we are in the presence of the triumph of the object. Pop art is precisely the transposition of the triumph of Venus, and Venus has become industrial plastic. At the same time we have the radical defeat of the artist's creativity. To put one's signature on a can of preserves or to add a touch of paint on a jar of cocoa, does not make these objects artistic; it is to admit that art has become completely impotent as a symbolic representation of the world and that it abdicates its responsibility by choosing the plastic lighter or the paper plate as the sole object worthy of attention, of admiration, of veneration, for modern man.

Now, this stance also has been radically challenged: in the face of the flood of objects it is no longer the object that one must conquer, as did the impressionists or abstract artists. Experience is built on the triumph of objects, which one cannot escape. As soon as anything is produced it is inserted into the world of consumerism and the all-consuming presence of objects. The work of art, when it is produced, becomes, in turn, an object categorized among other objects like the Frigidaire and the inflatable armchair. And so, one must simply break out of this mindset. And this can be done in two ways: by mockery, where the sculptor would place a plastic doll in the arms of Aristide Maillol's nymph, or he could add a carnival hat to Jean-Baptiste Carpeaux's *La Danse*. But mockery is nothing but despair without meaning for the public. Mockery is essentially aristocratic and in the end is ineffectual and lacks aim.

Jacquart is correct to speak of the theater of mockery beginning with Antonin Artaud, Beckett, Ionesco, and Arthur Adamov, and with Ionesco's manifesto, but this mockery constitutes simultaneously a critical message and a search for a unique event, a "pure" drama. "The notion of purity is the key notion, and one finds it in music and in abstract painting."[11] Purity can be obtained through *abstraction*, through the *unique* event, or through *specificity*. But this purity cannot exist if there is a message to be conveyed. Purity can only occur if we give into a pure form without meaning and ambiguity. Beckett and Ionesco are at the turning points of this breakthrough or at the rupture between these two orientations.

Other than mockery, we could simply refuse to continue the production of objects as in *the happening* [translator's emphasis]. No more text, which is itself an object, no more preparation, no more *mise en scène*, no preconditions or set. What happens happens. No more score, script, recordings, camera, or tape recorder, all of which are objects. And the sound track and the film stock are also objects. And the musical score is also an object. Ultimately, the artist who consciously struggles against this inundation of objects, who refuses reification, who has understood that everything is recycled not so much by society as by the universe of objects, condemns himself to silence and to making essentially nothing but an instantaneous gesture leaving no trace behind but a flash of life in a universe of things.[12] But this is not a creative flash of inspiration, because it leaves simply a new thing as witness of this occurrence. We must abandon the very idea of a finished work. There is the famous "work" of John Cage entitled 4'33": a human mannequin is seated by a piano on a concert stage; for four minutes and thirty-three seconds no sound is produced, and then it is finished. The "de-composition" is complete. But this can only be acceptable, let us note, from a creator well known, furthermore, for his outstanding, established, and performed works. The silence, itself, is a desperate abdication; it yields the field to objects. After the violent attack against the object, here we are again at its triumph, and with the most serious, most profound theoreticians of modern art. What a triumph!

One is plunged into anxiety when one sees a painter as prodigiously talented as Marcel Duchamp, neutered by the need for the negativity and the mockery produced by the milieu in which he finds himself. What else could it be but mockery toward art, society, the spectator, the connoisseur, in this perfectly mechanical and conformist world. The irony and insult expressed in the boxes, the banal objects, the montages. The sterilization by the will to not be reclaimed, to not become one more piece of the world being built by technique, and thus, radical impotence, while at the same time subjecting technique to the least questioning. Or, going back to the *Atlas Eclipticalis* of John Cage, for "a piece for one to ninety-eight orchestral musicians, 25,000 sounds roaming freely for two hours and forty minutes." This does not measure up to traditional music (of the Western world since the fifteenth century!) as I judge it here nor to the function of music in all societies; hence we abandon the finished drawing in order to de-situate "organized-disorganized" sound and to deprive it of all relation to our society. Unfortunately, this effort at rupture through mockery ultimately challenges nothing and leaves the artist's initiative neither here nor there in the ocean of whatever (*n'importe quoi*).

Indeed, we hear again and again: art is nothing more than the producer of objects. And, what's more, art is merely the creator of objects. You might believe that art was meant to convey something. Absolutely not! There is no sense, no subject, no history, no theme, no expression; there is nothing "to be said." There is no signifier nor signified. "Only the object is there." That's all. The novel is an object. "It is the text that one reads, the reason for the text and not, as if it were privileged by various received ideas, in the ebb and flow of the extra-textual. One can therefore wonder if the interest directed exclusively to the book of life does not have as its role to bring about a perfect occultation: namely, the life of the book." (Jean Ricardou) There is no longer a great glorification of consumer society and of the universe of objects. Art's sole purpose is to produce objects (Nikolai Tarabukin, *op. cit.*). That is to say, it does not have to produce existing objects and, in particular, things of the natural world, but it is a producer. It fulfills, ultimately, the same function as industry. It creates new things. We are less in the presence of a "constructivist" school than in the presence of a "productivist" school. (One cannot understand why the school in question and Tarabukin were condemned in the

USSR! What simplemindedness!) This simply reflects the technical mentality itself and not the reality of technological things. But, at the same time, one must also be aware that art is devalued amid this invasion of objects: it seems completely evident that an "art object" is placed among thousands of others and is no longer a focus of attention or contemplation. The artist is no longer a creator *vis à vis* his prodigious creations of products, materials, utilities, needs, signs, spewed out every day by technical operations. All creativity is concentrated in technique, and the millions of technical objects attest to this creativity that is so much more dazzling than all that painters or musicians have produced. (Hence this rage that sometimes seizes an artist striving involuntarily to produce as fast as machines—one, ten, works of art in a day.) It is as if this artist were placed in the impossible position of creating on the fringe of society, as if he were standing on the bank of a giant technical current. In society as a whole, this will translate into a movement of intellectual activity to a second degree, to that of reflexivity. We will see a contemplation that, at first, explains but then finally duplicates itself indefinitely. We notice this with the issues of modern hermeneutics and with structuralism. Hence, this increase of reflexivity that has become passive and sterile artistically and intellectually because of the excesses of technique. Artists and intellectuals, reduced to non-creativity in this situation, find a justification in this reductive reflexivity, and the circle is closed: and so they are placed exactly where the technical system wants them.

Under the influence of the technical phenomenon the object becomes the only important thing, all that matters. The search for meaning, for beauty, for communication, for values, for moral ideas, for metaphysical enquiry, prevents seeing the painting or hearing the music, because it is painting "in-itself" and music "in-itself" that is the important "thing," the painted object, and there is nothing else to search for behind it. One can write about faces painted by Manet: "Do not ask if they convey a feeling or an idea. They share the calm of a still life (*nature morte*). They are splotches of paint whose only interest derives from the technical issues they raise." This is even more indisputably true of the forest people painted by Rousseau. Is this primitivism? Perhaps! But then it is a primitivism of the technical era that belongs to the universe of objects where man is not distinguished from other objects. It's a swatch of paint, no more, no less. It's the reduction of the human face to an object among things; such is one of the great works of modern painting. It confirms and expresses nothing more than the reality of technical society. An extraordinary claim from Francastel in an interview for *Le Monde* is revealing: "Many intelligent and cultivated people are 'blind' in front of a work of art. They are in the museum: what do they see? They see the 'name plates' that inscribe the title. They see the subjects. They don't see the pictures. An artistic sensibility is struck, not by what the things represent, but by what the pictures are. As for me, I never pay attention to the subject of a picture. Never! That does not interest me. I begin by looking at the painting. But, for a painter, the reaction is even stronger. A painter could not care less about knowing what a picture represents. For him, there are good and bad pictures, which express a coherent totality of observations." (*Les grilles du temps,* March 1977).

Painting is no longer an escape to the beyond: one must concentrate on the surface of the canvas (moreover, one must refuse the fiction of perspective): we are led back to the fact that painting is simply a "painting of painting," painting as object, and no longer a springboard for thought, a symbol of infinity. We see to what degree modern art, which Malraux desperately seeks to integrate (despite the tragic admission to Vasarely that painting is finished) is the contradiction of everything that Malraux has claimed about art. And he seems to have proclaimed this at the precise point when only objects have meaning. This is a desperate perspective. The *text* in itself, likewise, becomes an object and as such is "sacred." This text should be submitted to a test of multiple readings, in a network of analysis with the attention of a scholar of the Kabbalah or of a theologian examining a sacred text, or with the attention of a scientist in relation to the objects of his study. The text is taken as an object. The writer writing becomes an object. The objectified novel treats everything as an object. This is hard for us to understand, but it is just as difficult to believe that all this is solely the product of a creative artist who is the complete master of his creation. It is no accident if all of this is just the process of reification! We know that music follows the same path to produce objects made of sound. Music can only be composed with "packets of sound." Moles reminds us that Meyer-Eppler sees authentic composition as the attitude of the musician who assembles according to his euphoric fantasy *all the sonorous objects* drawn from the surrounding universe or from electronic possibilities. "The sonorous object is a sound, more or less complex, more or less long, and so forth, defined physically in three dimensions (time, frequency, intensity) and defined subjectively by a certain number of qualifiers (a fluted, cellular, harmonic, or gradually intensifying sound)." But therefore the reference to an object is important in all cases, and, whatever the art, ultimately, it is the produced object that is important. Several years ago Charbonneau wrote, threateningly: "Art seems, once again, to be engaging itself in creating an illusion of reality, but let us be careful; if it is no longer a question of God, it is now a question of painting and not of reality …" Enough said. But, the modern artist would respond: "Of course, the *only* reality is my painting or my musical composition." And this dominance of the object agrees with the will to separate the text from its meaning (*sens*) and from the one who speaks or writes. Language, like art, is separated from intentional meaning because the object is only an object. Francastel still, in the same interview, declares that meaning does not account for the work of art, that the work overflows all meaning, and that artists are not interested in what things represent. "They have a direct perception of the outer world, which is definitely not conveyed by the word, and they convey this perception in like manner." And for illustration, he evokes Fernand Léger, Matisse, Robert Delaunay. And this is perfectly correct. But one must remember that this is the modern condition. It has not always been this way. Rembrandt explicitly sought to say *something*, to transmit a message, to witness for his faith. All else was a means, a more or less adequate support, more or less truthful. Clearly, if one produces an object as such, why would one wish that that object had the least intellectual or emotional meaning? That object can speak through itself but only of itself if it is separated, as modern art demands, from its creator. Roland Barthes has forcefully stated this (especially in *Sade, Fourier, Loyola*, 1971). Writing as pure object is more meaningful than the content of the text. It is not, therefore, correct to state, as above, that there is *no* meaning but rather that the meaning is independent from the explicit message and from what the author intended and also from the "background" of the work of art. On one hand, it is quite evident that the work of art surpasses or overflows the meaning (*le sens*) and the intended authorial message, if any, but on the other hand, is it right to cling to that object alone, which overflows meaning, while belittling the intended or intellectual meaning of the sermonizing or of the pretext for the work? That's what leads back to the object itself: one does not need to know who has painted, who has composed, or who has written, nor what were their intentions or convictions. The object is simply there. There is no longer a story nor a meaning. One must surrender to the object. That's all. What is surprising is that the artist, deeply subjugated to this deep urge in our society, renounces, in turn, all attempts to say anything. That's why, as we said before, parenthetically, "if he had the will." Now the creator turns out objects. He does not wish to express anything. This simplifies the situation. The interpreter and the creator are on the same page. But, why is it this way?

All the elements that we have just pointed out correspond perfectly to the technical process. The reduction of the work of art to an object is identical to the reduction of traditional craftsmanship to the blind output of mass-produced objects. The rupture between the worker and his product, the disappearance of the personal mark or touch, and above all, the disappearance of meaning or sense, are all a direct outcome of technique. How could the object be considered other than in-itself when it is mass-produced by an enormous industrial process? It is clear that technique supplies "these objects," and that is all. The origin, the intention, are missing, as well as the meaning, because technique, alone, eliminates all meaning. Technique leads back to an operation, whether it be the producer of objects or of abstract procedures. There is nothing but production or product. In no way is there either an objective or a finality.[13] There is nothing but the process of function to which the work of art is now completely assimilated. The extremely rapid obsolescence of all that surrounds us negates the work of art, and this process mirrors perfectly the production of technical objects. Conil Lacoste might have had this in mind (*Le Monde*, October 1967) when he wrote:

> *For a long time, now, one could speculate on the probability that the weirdest output of the avant-garde—metal mobiles, collages, garbage cans, burnt nylons—will still be recognizable in half a century as works of art, when our grandchildren discover them among the junk in an attic or the hurly-bury clutter of a basement, assuming these things are still intact.*
>
> *What kind of longevity or hope of cultural life can one assign to metal casings and flushing mechanisms touched up by Tinguely, and the pieces of metal rods by Viseau, to the "compressions" of the excellent César, or the blue sponges of the celebrated Yves Klein, whatever their justification in a given context and however justifiable our adherence to them when the sublimation of the ingredients in question sufficiently surpasses the desire to astonish? And how many of these outrageous products have fallen into oblivion through the absent-mindedness of a truck driver or the ignorance of a customs officer? One could go on and on.*

This is the fate of millions of products of the technological society that one throws away after having used them once (wrapping for example!) or as soon as they are a little stale. The triumph of the object is the quick obsolescence of the object. And for art it is the same as any other product. Art has aimed since its origins at an eternity, a permanence. Now it has changed utterly: henceforth it has been objectified, that is to say, it leaves the sphere of the eternal, of being. And this is true even for art that "carries a message," which is always about a current event, which is as quickly forgotten as the last radio broadcast. The painter scrupulously delineates on his canvas the latest headlines about the war in Algeria. The playwright gives a traumatic account of the war in Vietnam. But once these events have passed, nothing remains. Thus, expressionism itself is a product of the technical society. Expressionism, as Adorno demonstrates, contrary to belief, is also an example of an objectification. The content of expressionism, the pure subject, is not pure in its apparent isolation from the technological society. Once again, it's a matter, quite simply, in all modern art of an allegory of technical society, but an allegory that completely adopts the traits and orientations of this society. We must go further. At a certain level of production, universal technique continually creates new objects and ends up by producing what is perfectly superfluous, useless, futile; thus, the most efficient, rigorous, and rational system ultimately creates stupidity and incoherence. In like manner this new objectivity falls into what it most strongly rejects: ornamentation and the baroque. Everything in a work of art that does not have a determined function (and it is clearly the structuralist project to unearth the functions in every text) must be eliminated. Rhetoric, decorative touches, and baroque elements, are to be entirely eliminated, but what is totally rational ends in irrationality;[14] the totally functional work of art is completely bereft of function! Hence, we fall back into what is most grotesque, *Kitsch*.[15] *Kitsch* denies what is non functional in the work of art, what is not reducible to the object, because it is this non-functionality that transcends the simple existence of the object's *being there*. And, "as the work of art cannot be reality," as Adorno says, "the elimination of all traces of appearance can only underscore most flagrantly the character of appearance in its existence." But we can go a step further: for the work of art, as we have seen, claims precisely to be reality, just as the technical product claims to be a new milieu.

The direction taken by Dubuffet is a good example.[16] This extraordinary painter, who has used so many procedures, subordinates them to a fundamental critique, and it is the tenor of this critique that is significant. His thought, his research, corresponds to the truest life experience of a man submerged in the technical milieu, and his approach is actually that of technique in relation to the natural milieu. He seeks not only to free himself from all traditional painting but to critique the entire culture, the "spirit of culture," and, what is more, the milieu in which man has always lived. Dubuffet is going to "dislodge" appearances with Raw Art that negates both object and material. Anything goes: paint emulsions, plaster, mud, lacerated paper, nicks, incisions, "caustic attacks, puncturing and pricking." Production and dissolution, figuration and defiguration, must occur in a single movement to dive into the interior of matter and objects in order to deny and to dissolve them. All construction is abolished through the outer surface. He plunges into the "unnamable" with streams of paint, asphalt, filth, confusion, and the formless. He dislodges as matter itself is dislodged. He detaches himself from form; raw matter is his arena of expression but at its lowest level of throwaways, slag heaps, and garbage. He nullifies man as an object and the object as matter and dislodges matter itself. It is the disintegration of the old world by means of the technical implosion. But immediately we have a question: "Why not let matter invent its own forms?" Hence, after the dissolution, the dislocation, there inevitably comes, in a second stage, the "invention of materials." That is, painting, stripped of itself, obeys matter and the materials which take over and create. And here with the most artificial and synthetic ingredients and other sophisticated chemical products, the painter recreates matter out of raw matter; "misleadingly, they imitate the pebbles, the earth, the natural substances, none of which is entirely artificial nor the product of chemical processes. The material returns, but only in appearance, manufactured by the painter, and thus, the inexorable object! The whole direction of Dubuffet's work (which we will not follow in its third and fourth periods)[17] expresses perfectly the pretension of technique, typically demiurgic, in the very process of reproducing nature and matter, by artificial procedures, after having dissolved that material into its component elements, after having expelled the material object, creating a new world through millions of objects, as Dubuffet produces furiously those formless objects, one or sometimes several per day!

We, therefore, have a series of assimilations between modern art and the technical system, which is situated at another level from that which we had mentioned as clichés. In the same way as technique, the artist reduces his understanding to objects. Michel Butor explains it clearly with reference to *Description de San Marco* or in *6 810 000 litres d'eau par seconde*; words are pebbles, of which a given ordering will restore for the reader the Basilica of San Marco, not just by a description; rather the words are drops of water whose collusion creates the movement of Niagara Falls. The words are anything but carriers of meaning, and here we find, marvelously, the equivalent of a great technical process: thanks to such a magical operation, one will succeed in producing artificially what has existed and has been produced for millennia: the shimmering of particles, for example. It is astonishing. Butor, by the same token, recreates San Marco for you with words that give not just the impression he has had of it nor any meditations it produced … all of that is commonplace and banal. No, it is San Marco itself. But the Basilica exists. Why re-create it? The only meaning is in the prestige of the magical technical operation with the disappearance of the subject. Butor is an excellent technician who thoroughly participates in the technical system, only that and nothing more. Here again we have in large part gone beyond the direct influence of technique. De Chirico and Léger painted elements of technique and painted about technique, expressing the collision of the human with the technical. We are beyond that. There is no longer anything human, no longer any drama about "future shock." The artist has entered into the game and translates in his work the essence of technique. He paints, he writes, not on the subject of technique, but rather his work is the profound expression of technique to the extent he is conscious of what he is doing. He does not know technique. The barometer does not discourse on atmospheric pressure; it knows nothing about this. It simply translates it—that is all, just like the truly modern artist. He cannot do otherwise because technique is basically the world in which he lives. And so, to end with a final example, René Curring works on recording shortwave broadcasts from all over the world. His work is a matter of sound combinations, of cracklings, of indistinct words, of fuzziness, of explosions, of squeaks, of yacks. That is, he reproduces what one actually hears when one twiddles the dial in the shortwave zone. Is there any characteristic of the "work of art" here? Is there a will to produce, to arrange, to order, according to rules, according to musical laws, according to a canon? One does not know. Thus, the basic sound is not the song of birds or the sound of the sea. It is a mechanical noise produced and recorded by a mechanical device. Man expresses this milieu but he produces nothing but mechanical noise. He expresses a technique in motion. He is, himself, a technician transposing that technique. He is not touched in his being or heart by his relation to technique. He is situated exclusively at the level of his own adaptation to the technical system. He does not transmit anything seductive, "beautiful," agreeable, or sensitive. There is no view toward taking stock of the horror of the universe of objects and of technical mechanisms, no indeed! That is no longer in play. He transmits simply a faithful interpretation of this universe that is thereby given to us, and it is not ours to question. That is the fact. Art now expresses objects and the power of things.

Thus art becomes the reflection of the technical reality, but, like a mirror that reflects an image, art does not know its reflections as images. Art limits itself to being an indicator. Painters, film-makers, musicians, sculptors, understand absolutely nothing about this technical system. When they discourse on it, they sidestep the issue, but they are, in spite of themselves, the effective proponents of this art, while, at the same time, evading the reality of technique. There is a double mechanism for this process of evasion (and this corresponds to the contradictions noted above). Sometimes this art speaks of another reality essentially political, the Marxist conception, for example, and fixes our attention on these red balloons, by turning our attention away from the truly real. By classifying the political and revolutionary problems as the only decisive ones, art turns us away from real problems, which exist in a concrete situation of our time: art creates chimeras and fantasies. Sometimes art tries to set us up inside a purity, an icy theory, a surreality, a pure form, and this leads us to an "absolute" that is a simple evasion. Nevertheless, we need to consider, as occasionally we have seen, that those artists attempt to account for the industrial world with concrete music, anticipatory cinema, and so forth. Are they not the ones who cause meaning to spring from this world for all to see? In reality, this is not the case, because a two-fold reaction occurs: for the specialist, the technique of these works is his consideration, and nothing else. For the public, there are responses of unease and distaste: concrete music is unappealing; *Mondo Cane* is odious, and the reaction might be: "This is not what I experience! I can more easily accept what I experience instead of those images and sounds that are not true but are hateful." And when all is said and done, the spectator, the listener, is led to accept the life that is presented to them by the repellence of that image, which they reject. The technical world is not hell. It is more subtle. The image of hell that we are forced to accept with gratitude is the technical world. Thus, sometimes art may carry a compensatory message for the intolerable nature of the technical world, and, sometimes, it mimics technique itself, but it is always situated in relation to technique and simply carries out its role forcing conformity in all its schools and expressions.

Notes

1 Now, in a confused way, what we express here has already been foreshadowed by Mondrian. In his *Manifeste néo-plasticiste*, there is indeed the proclamation of a denaturalization of art in order to obey the new world of man, the industrial world. For man and for the artist it comes down to forgetting the "quaintness of nature," and the "roughness of the rustic." The natural landscape must leave the landscape of art. The "nature" that "the artist must model is new mechanical space." This, which has been called industrial romanticism, relates more to a proclamation than to an analysis; but the intent is clear. And all modern artists, whether they intended to or not, have followed this path.

2 What is significant is not in the *direct* imitation: the gearwheels of Léger, the *Electricity Fairy* of Dufy, the *Air, Iron, Water* of Delaunay do not appear to me any more definitive than the machines and profiles of gadgets of Millecamps, today. This imitative pseudo-symbolization is not the true testimony of technical influence.

3 For one in-depth study of the phenomenon see J. Ellul, *Le système technicien*.

4 Of course Kierkegaard had already admirably understood this in his analysis of the aesthetic stage. But this has been horribly generalized and stressed: a good example is the novel of John Hawkes, *The Blood Oranges*.

5 Rookmaaker cogently remarks on this subject: "In minimalist art, one uses forms inspired by metal factory pipes or building girders. These objects have an amazing beauty in their own environment fulfilling a specific task, but they become strange and hallucinatory when one takes away their sense of meaning isolated from their context."

6 Jacques Ellul, *The New Demons*.

7 These lines were written at the same time as Roger Caillois's astute study on "Picasso the Liquidator" appeared in *Le Monde* of November 1975, which scandalized and which broke so fortuitously with the beatifying discourse of unconditional admiration for Picasso, and with which I agree entirely. I recall the main points: with Picasso there is a new kind of disrespect, a disrespect for the order to which he himself belongs. He is irresponsible, and his success derives from the fact that his painting does not aim at consequence. Picasso is arbitrary and derisory (terms that Caillois borrows from Malraux's *La tête d'obsidienne*) and in this respect the work of Picasso is completely relevant, since, as we have shown, all current art is quite arbitrary (except in regard to the technical imperative!) and derisory (in regard to all except the new sacred!) And Caillois hits the mark incidentally when he says that, while the forms created in nature are the results of an infinitely slow process, those of technique are in constant flux. What we term Picasso's constant inventiveness, is no more than following technique's rhythm. Hence, Picasso is beyond any possible artistic following or tradition. "In no way do I see him as the prolific sower of seeds for the future but as the canny and sardonic liquidator of a century's old enterprise, the dissolution of which he anticipated, like the rats fleeing a sinking ship, he hastened this final chapter through his lucid speculations. All of Caillois's argumentation that I have not reproduced here is of a precise rigor and exactitude (that his detractors have not grasped) that one must read and reread this admirable study.

8 J.-J. Goux, *Les iconoclastes*, 1978.

9 It would be necessary to make a special study of Beaubourg [Centre Georges Pompidou] simply as the expression of the technicization of culture, with its costs (one billion francs for its construction, 131 million francs for maintenance, the cost of more than five universities). Its energy consumption is equivalent to a city of 10,000 inhabitants, with the strangulation of other cultural forms (given its enormous costs it was necessary to reduce other subsidies) and with the concentration at one point of all possibilities. Because this is the monumental mistake: to manufacture an enormous culture machine in the center of Paris, bleeding the rest of the country dry. Assuredly, as a machine for communication, this works, it sells, and brings in the crowds, but who are they? Does one believe that what is produced as a great hoax can truly produce cultural development in the world of the worker or the average French citizen? This is why the official discourse, which claimed that Beaubourg would be a "center of artistic creation," is perfectly delirious. Neither those from the country nor those from the world of labor would fall for anything in this pompous circus. Once again we have the coagulation of the specialists of culture, on the one hand, cheerleaders for design, for acoustics, for avant-gardism, and on the other hand, swarms of minor artists and intellectuals in their stalls selling, exposing, and publicizing themselves to no end, and they will make up for all their deficiencies with catalogues, computers, videos, microfilms. It is an enormous self-gratification machine concocted by politicians and technocrats of information culture in which prestige and the flow of money are essential motivations. This said, one must specify basic principles that preside over the creation of this kit and caboodle:

1. The state has taken over culture.
2. *The more money, the more culture*: culture is expensive, and one must not skimp on the means: for a hundred billion old francs one will get a lot of culture (incidentally, the money-culture relationship corresponds closely to the image of luxury culture).
3. *Culture is necessarily non-conformist* and even revolutionary (hence the officials of the Centre Beaubourg must manifest their political and cultural opposition to the government.
4. Nevertheless, *culture must be tightly integrated into the actual world*, hence, technique (which, on the one hand, corresponds to the Center of Industrial Creation, design of products, data banks on consumerism, etc. etc. and, on the other hand, the maximal mobilization of every pseudo cultural gadget).
5. As one knows, since artistic creation has been sterilized among the people, culture can only be created at the grass roots, hence the offering to the people of libraries, museums, theatres, studios, where anyone can come and create his own culture. Discourse overtakes reality.
6. *The more technological apparatuses, the more imaginative possibilities* (a fundamental paradox, everyone knows that a child has a more intense imaginative experience with a bottle cork than with an ultra sophisticated

gadget. Beaubourg is the cultural showpiece of a sterile, technical society.

Bibliography:

C. Mollard, *L'enjeu du Centre Pompidou*, 1976 (*very conformist; much more interesting:*) G. Affeulpin, *La soi-disant utopie du Centre Beaubourg*, 1976; Jean Baudrillard, *L'effet Beaubourg*, 1977; A. Cauquelin, "Le vide Beaubourg," in *Revue Projet*, 1977.

10 Just a word on this misunderstanding: an article from *Le Monde*, July 1977 notes: "Luxury and progress, at last, *within everyone's reach*." It is about a wonderful new automobile that brings together all the latest advances and aesthetic accessories. The article insists on the democratization that such a product brings. That is all well and good. One small detail: the price—65 thousand new francs—in other words the equivalent of three years' salary of an autoworker. Some democratization!

11 One must distinguish, along with Jacquart, between the theatrer of the absurd and the theater of mockery. Sartre and Camus held forth on the absurd, producing classical and traditional plays, in order to make us reflect on the notion of the world's absurdity. Beginning with Adamov and Ionesco we are plunged into the absurd through mockery, and we are given the physical sensation of the absurd and made to experience it directly.

12 This refusal can assume a highly elaborated technical expression: thus the well known homage in New York to Tinguely (1960). "A work of art constructs and destructs itself on its own." The work was an assembly of pieces producing effects of painting and sound that was to self-destruct.

13 On the total absence of finality in technique, see my *The Technical System*, *op. cit.*

14 Concerning the rational and irrational in technique, *cf.* B. Charbonneau, *Le système et le chaos*.

15 G. Dorfles, *Il Kitsch* (ed. Ital., 1968). *Kitsch* is a German word untranslatable into French that is accessible to the general public, based on reproduction. *Kitsch* goes hand in hand with advertising and fashion. *Kitsch*—cemetery statues, decorative grills on the Metro, shops selling cheap, religious imagery, and also shops in the Latin Quarter that sell oriental, Chinese, Hindu, Vietnamese, and Eskimo works of art. *Kitsch* corresponds to a "fashionable bad taste identified with the vacuousness of advertising." *Kitsch* possesses the external forms of art without any creativity. The computer can only produce *Kitsch* for this occurs only through the great quantity and the repetition of forms that were once original.

16 Max Loreau, *Jean Dubuffet. Stratégie de la création*, 1977.

17 All this will be explained in a detailed study soon to appear.

Chapter 3
The Message and its Compensation

Here we approach the first face of the aesthetic Janus, which looks in two directions. Modern art is a means of action, a lever and a bearer of a message. It is a self-affirming ideology along with its denial: in relation to the milieu it claims an aesthetics of play and of escape, avenues closely intertwined. Escape claims to be revolutionary whereas a message is a compensation.

I / Technique engenders an Ideological Art

Let us avoid a misunderstanding: by "ideology" we do not suggest the cliché that has been repeated a hundred thousand times and held to be a scientific truth, mainly that: "Everything is politics, everything is politics of class. All art expresses an ideology of the dominant class, where the dominant ideas are those of the dominant class, etc." Pseudo Marxism has no value in any domain. We should consider facts like these: how does it happen that Soviet art from 1925 to 1960 is perfectly identical to the official Western art of the nineteenth century; how does it happen that, since 1965, all of the ideology expressed by Western art, including American art, is exclusively revolutionary and confrontational? Can we say that in the USSR there was no socialist revolution or, likewise, that bourgeois ideology is no longer dominant in the Western world since 1955? This is quite complicated. How does it happen that even after 1960 in the USSR the evolution of the ideology of art follows quite closely with that which was produced since 1890 in the Western world? Indeed, one must go back to the Marxist interpretation of ideology as a reflection, a veil, and a justification, as an aspect of social reality and not simply an expression of "class domination." This new reality that is technical society (which also involves relations of domination that are no longer those of the nineteenth century) is an ideology that serves both to veil and to justify with the aid of the arts. We will go beyond this nineteenth century Western ideology and will add compensation to Marx's triplicity. Art is, in effect, a phenomenon of compensation in relation to the real. We will come back to this. But first we must consider the growth of the ideological factor in art, which has greatly mutated, a phenomenon not unique in the course of history.

Art has always functioned in the process of symbolization, but as we noted in the preceding chapter, the technical milieu wars against symbolization: ideology increases as symbolization decreases. Ideology is introduced consciously and willingly to give meaning to an art that ultimately loses its symbolic force, which is replaced by ideology. Thus, from the intentional side of ideology (the revolutionary message, for example) we have what the artist wishes to express, and then, from another side, we have an unintended ideological role that flows from artistic expression. For example, we have two strategies:

to veil or to justify the technical reality or to compensate for what is intolerable in the technical milieu by allowing the citizen to decompress and unwind. But the two phenomena of symbolizing and ideologizing do not play at the same level. Symbolization clearly addresses ultimate problems like the meaning of the world and of life, the problem of an eternal that is either present or absent, or the problem of birth and origins and of death and final ends, while ideology confronts problems more nearby and immediate. The substitution of the ideological for the symbolic is one aspect of the so-called Copernican revolution (multiple and successive): to bring back everything to the human world and to center all on a current political life. There is no elsewhere, no meaning, no great beyond, no eternal; there is only the human in an immediate, temporal life, in an *in-sistence and not in his ex-istence*, in a social nature: all of which must be considered in an ongoing manner, in a political reality. This "Copernican Revolution" is, in large part, the work of *technique* or, rather, of the entire technicization of society. Symbolization has fewer and fewer opportunities to exercise itself in a human universe exclusively restricted to the person alone, to the master and measure of all things produced by and in technicization. Equally, technicizing entails ideologizing. The over-powering adventure with *technique* requires ideology. Such considerable effort would be impossible without ideological motivation. Further, a system so demanding and rigorous, so far beyond human experience, begs an ideology that hides the reality of the system, while at the same time lending support for it and for its consequences. Hence, the more the system is rigorous, rigid, and totalitarian—in a word inhuman—the more it must conceal the actual reality of the ideology with compensations that make the situation tolerable (the ultimate costs of the automobile—city crowding, all manner of noise and pollution that remade cities—is veiled and compensated for by the passion for speed, by the manifestation of social prestige, etc.). However, this ideology reveals, at the same time, man's essential denial of this reality. Man withdraws either from his own life or from his former life, or from politics, or even from an aspect of the technical system. There is a fundamental refusal that is expressed solely in an ideological way that plays itself out at this level. Here, compensation enters: one mounts an ideological attack on the system in question with the result that, so satisfied, one goes no further. The ideology is self-justifying to the precise degree that it raises questions, but this is a justification in the second degree to the extent that it empties out and exhausts the human capacities to resist the inhuman. Thus, *in this sense and with regard to the system* (I don't say always and in every case), ideology by necessity raises questions. But this refusal, this questioning, is precisely what allows man to justify himself, and, at the same time, to become part of the system, and, consequently, to support it. Man seeks to prove that he has a reason to hurl himself along this path into the technical adventure, into the working out of *technique*, which he proves by asserting his freedom. How better to be free than by peppering his world with criticism, by taking a challenger's stance and raising questions. This is his little combat medal: the "proof that I'm not under the heel of the system is that I am a destroyer of morality and taboos, that I am a terrific revolutionary, that I insult my parents and the president, etc." What appears as scandalous is really a marker of success.

Everything gets used up quickly, and the reaction of society is never effective against those who would provoke a real upheaval. We have two hypotheses: either the upheaval is fictional, because it affects nothing important, or else, society has completely disintegrated. A characteristic example is the Rolling Stones: scoundrels insulting the public, accumulating misdemeanors, forsaking music for hellish sounds, drugs, etc. We have a few vague reactions in England, enthusiastic reception from the youth who believe in transgression, enormous receipts, the use of violence and insults to increase attendance, a total absence of any message whatsoever except for the triumph of the almighty dollar, and then, a stamp of approval from good society, of course. The Rolling Stones in 1975 are a spectacle for cultivated people who want to be with it. They offer the public false violence, false transgression, false eroticism, which is just what provides the compensation for living in the technical world. Thanks to gadgets cleverly applied, one has the impression of living like an Indian in a tricked out Wild West or of being caught in a hurricane or playing cops and robbers. One pays a price in order to join this ostensible attack of the world as it is. But one carefully avoids questioning the technical system! It is formidable only because it is uncontested, but it is precisely the ideology of contestation that allows one to show that one is in no way socially determined. *Technique*, however, is never called to question. Ideology plays its role by attributing the evil of man to everything and anything (exploitation, class warfare, morality, faulty upbringing, the Oedipus Complex, etc.) except … to *technique*, which is certainly not untouchable but is simply forgotten. And when called to question virulent reactions are immediately provoked: the ideology of contestation does not permit self-examination. Ideology would become an un-veiling instead of a veil, and, hence, it would cease to be ideology. But man holds on to it. He holds tight to his ideologies. Ideology is a compensation for the real in the sense that, by not attacking the reality of ideology but rather by making what is unreal appear real, ideology provides the feeling and the conviction of power, of autonomy, of lucidity, of action, etc., which do not exist in the real world but only in the world of appearance. Now this entire schematic set forth is characteristic of modern art. It is ideological in that it is the privileged instrument of this ideology. It contests and compensates, it hoodwinks (that is, it focuses attention on false problems); it passes itself off as reality (but a false reality). It is a process of human justification that is never questioned except through art. In other words, man calls everything into question except what is essential (what is essential is no longer God or morality but rather *technique* and power). Art is not only the bearer of this ideology, the expression of the veil and compensation, it is *itself* this veil and this compensation; it is nothing more than ideology.

Hence, this is all the more remarkable in that art has a privileged place in the modern ideological process, which is itself an essential piece of the technical system. Indeed, art, as we have shown, exists only thanks to modern *technique*, under all its aspects: materials, procedures, but also its mentality, etc. It is *technique* itself utilizing technical means; it is the foreground of the technical system; it is formed, for example, out of the mass media and it is, simultaneously, the ideology, veil and compensation. It produces a radical illusion, all the more subtle because it propagates what it hides like a shell game in which a more technicized art becomes more confrontational, more outlandish, as in the work of fringe artists and the great outcasts who do no hesitate to call All and Everything into question. One can proclaim, therefore, that the artist has a social function. In standard ideology the artist is the great rebel. He must "form a bond with the struggling people: to exalt them, to depict them, to love them, to attempt to understand them, to talk to them, to talk about them, to show them as they are." (Alejo Carpentier). This is the veil. But in effect his social function is to be an entertainer who distracts from the real by his very violence or his ingenuity, and who manufactures, thanks to *technique* itself, an illusory universe that permits man to continue tolerating the frightful condition to which he is reduced. Later on we will run into an aspect of this when we deal with the artist as archetype of the free man. But the artist is caught in the contradictions that we referred to previously. Up until now art has always expressed the symbolizing function, which is no longer possible. We see a thoroughgoing repression that will lead to a baroque, exasperated, delirious artistic expression. Furthermore, the technical system, allowing for the imaginary museum or for putting art at the disposal of the mass media, places the artist in a totally new situation in relation to all that was both the meaning and condition of art. Leaving behind a past that is no longer possible is inevitable, but it is also inevitable to flee this perfectly intolerable and unacceptable situation. To escape, the modern artist will delve into the ideologies that surround him—social, political, and metaphysical ideologies—and attempt to express them. In the place of creative, artistic invention, the modern artist enters into a confrontational and ideological community that is the only possibility within the technical system. He will seek that which best fills the void, that is, what best allows modern man to experience the technical world. For example, it is at the exact moment when the landscape is covered with industrial discharge, when the city air is mephitic, when the atmosphere is a haze, when one approaches what will be called pollution, that one exalts light, the true sun, and airiness, in a painting that becomes clearer and purer, thereby clearing the coal dust. There is no end to talk of light, which is a compensation for what is progressively removed by the industrial process. Light, which is delicate, diffused, and broken into infinite nuances, is the great theme of 1880. And when one denies that the painting is a "window" onto reality, one sees a picture opening to fictitious reality. The examples are myriad. And, at the same time, we will have an art refusing to be utilitarian, where an architect will imitate Gaudi or Philip Johnson by radically denying functionality. Philip Johnson, who defends the principles of an international style, explores the architectural domain of pure forms, seeking to put an end to structural architecture and producing pavilions that resemble those of de Chirico. There seems to be a contradiction between art refusing to be utilitarian by espousing gratuitousness and art that is ideological and politically engaged. This is not the case. The conflict is purely formal, apparent, and playing at a superficial level of stated convictions. In reality, the function of the two ideologies of art is the same: in both cases it is simply a question of denying the reality of the technical system and of concentrating the attention and interest of the spectator on something else. Only the approaches are different. It's the same expression yet again when art limits itself to being a passing outburst against technology, because it has resolved to be purely Ariel, not the more spiritual Ariel, but a fugitive Ariel. The Happening is an outstanding example of self-justifying and compensatory, but false, confrontation, allowing for *technique* to resume its function almost immediately. This is essentially the phenomenon of ["off," "off-off," "off-off-off,"] random performances, which were rampant in the United States at the beginning of the 1970s: challenges to the nasty business world, overtures to a purified art, expositions and spectacles spread by word of mouth, avoiding publicity to attack advertising, spur of the moment workshop demonstrations, unfinished inventions, stirring up scandals and provocations, spectacles … that surpass pornography because love is banished from the technical world, making nudity a spectacle because we are immersed in a media world of iron and electricity. But all that is nothing but compensation. Desperately, this compulsion of art wants to escape the contradiction in which it is trapped, where man is also trapped. It wants again to express the real and the true, but in so doing, it hides the truly real. It wants to express a truth of nature (nudity, love) and a reality (light, etc.), which refer to the past of things of the natural environment, things which have no place in the technical environment; these are nothing more than the backward glances of a false naturalism (but in a place where there is no longer nature in a natural environment or in a social environment). It is a return related to the innocent homosexuality of pastoral poetry, to Eskimo outfits on the Boulevard Saint Michel, and to the Afro hair cut, to the seeking of an origin as in this art itself (pure, unadorned, etc.), but all this remains perfectly inadequate, perfectly superfluous, like frills and furbelows on the rigidity of the system. One tries to express a reality in order to escape from the technical reality, a process that leads only to the people of the third world, or to the nature of the past. *Kitsch* is no different. So then, all that activity and hope collide with another reality, which is an unacceptable resistance, and one can say that the revolutionary aspect of ideology in modern art results from this very conflict. It is impossible to take account of the pure state of the technical system, but one wants to account for the real and the true. But the real and true are not what we encounter but rather what is imposed on us at every street corner. One must, then, challenge imposed reality. But since it is, at the same time, impossible to question *technique* itself, one hides a challenging and revolutionary ideology, which finds its satisfaction in the return to the Third World and to mother nature. We come away from this, and this is something we will have to deal with at length, believing that by changing political society, we will, at the same time, change man and the nature of art. But, by the same token, in the process of changing art, we are already changing society. However, we fail to understand the order and structure of this society. And the obfuscation is taken to its logical extreme by commentators on modern art, who, as precise interpreters of this art, are ardent defenders of *technique*. This is very significant. They understand to what degree modern art is the expression of *technique*, even when modern art pretends to be hostile to the machine, but the attack in this area does not get beyond the "machine"! Just as it is the product of *technique*, we can admire modern art to the same degree that we admire the *technique* that inspires it. Thus, Delevoy will sweep away "all literature without support," like Thomas Carlyle, John Ruskin, or Georges Bernanos, because it expresses infantile resentment against the machine. And Francastel sweeps away in the same fashion, but with condescending disdain, Mumford, Giedion, and Geddes, because they dare to question technical rationality. They are totally clear that the revolutionary nature of modern art deals with everything and anything except *technique*, which is wonderful progress. A subtle game develops where art challenges the superficial aspects of *technique*, for example the mechanization of life or rationality, or industrial culture, or false clarity, but instead promotes obscurity, symbolism (but in reality, pseudo-symbolism!), irrationality and subconscious drives. Nevertheless, this same art draws from the essential structure of that which it pretends to oppose, and becomes a contradiction to its own aim and message. The modern artist experiences more than anyone else the anguish of dehumanization by obsessively and joyfully revealing this anguish. A death wish? He seeks the most direct, the most unadorned, the least filtered expression of this anguish by rejoicing in the very negation of the concept of man (all contemporary painting and theater give witness to this), and consequently they become the mirror of the dehumanizing system that they claim to condemn. In fact, this art, no longer symbolic, becomes a ritual that exalts and incites the general process of *technique*. This art "frees" the drive toward the denial and madness of humanity drowning in technical rationality. But, in reality, it compensates and subjugates madness to the structured rationality of technical society. One needs to know very little about modern art in order to understand that there is nothing tragic in dehumanization. Even as dehumanization as an instrument of standardization ceases to shock, art becomes a spectacle tamed by madness itself and by unbridled excess. By doing this, art tries to escape *technique* by transmitting a *message*. It's no longer a matter of telling a story or proposing ideas but of provoking reactions that constitute the "message." Violent colors, film loops, distortions of reality, jolts of loud music, are all characteristics of underground cinema that is a kind of animated handwriting. Groups like "Change" or "Poetic Action" accuse others of being terrorists. However, a return to the people would permit an escape from *technique* and this dead end situation, but this return is either purely theoretical in its revolutionary discourse or else it's the product of a hyper-technical form of art that is perfectly hermetic. This debate is long running and is characteristic of technical society. Lenin, the hyper-technician, had already condemned the "*Proletkult*" movement, which was already heading down the wrong road. The "non-public" that Francis Jeanson describes could only be formed by an art that would be both accessible and non-technical and would attract only four or five percent of the workers.[1] Thus technical society produces an ideological art that justifies itself with provocation and protest but which can only allow compensatory expression for man's experience of inhumanity.

II / *Message and Revolution*

Modern art pretends to make us act or reflect. In New York, November 1969, a "show" of cool painting, called Street Events, organized by twenty painters to cover twenty city blocks in Manhattan to force passersby to see more clearly their surroundings and to make them aware of the real urban world they inhabit. The Paris Biennale presents itself as a gathering for promoters of ideas, for questioners of values, and challengers of dogmas, and here, again, we see an aim for awareness of the shattered lives humans lead. The Mexican painter José Clemente Orozco, in his celebrated canvas entitled *Academy*, shows brilliantly what the university has become, and the no less famous *Painting* (1946) by Francis Bacon starkly reveals the monster in which we live and which lives in us. But what is interesting in this canvas is that the point of departure was the painting of a bird, which, through successive deformations, has given birth to the horror of this world more strongly expressed by far than by the all too famous *Guernica*.

Now, this message, this vision conveyed by artists, is almost always pessimistic. Modern literature is fundamentally pessimistic,[2] and theater, whether it is Genet or Arrabal, or Dürrenmatt or Beckett, offers no hope, and painting or music are acts of unlimited, violent aggression. But, most often, the discourse contained in these works is, in reality, political rather than social. In spite of its philosophical title, Esslin's work, *The Theater of the Absurd* (1962) shows that the subject of all theatrical works since Brecht is political above all. The same is true of Max Frisch, or Günter Grass, or Václav Havel. The political nature of these works is sometimes disguised as fantasy or eroticism. The same analysis occurs in Franck Jotterand (*Le nouveau théâtre américain*, 1970), whether dealing with the living theater or Tennessee Williams (in spite of the apparent intimacy of the latter), be it Albee or Kazan: it is an indictment of society, in particular of American society, hence it is a political indictment; it is theater that addresses the new forces, "young students and blacks." Gilles Sandier, in his essay on *Théâtre et combat*, 1971, aims to show just one thing: only authors committed to revolution have any value. Ionesco is downgraded for his militant attitude. By contrast, Vauthier is brought to the fore. Peter Brook gives a very enlightening explanation for this. Theater should be a forum. He cites Us and the Iks, etc. and does not claim to act on politics alone; he situates himself between myth and anecdote, and thanks to the insertion of politics, he seeks to create an awareness of the universal reality of man. The same holds for the novel, which is worthy to the degree that it expresses the political reality in the broad sense of our age; the novelist is "face to face with the political world where at last one speaks in clear language." (Carpentier). This study is interesting because the author rejects an obscure and formalist request for a language that is understandable and is that of history itself that is created around the artist or author. Only political commitment allows the novelist to speak to the people. He observes, at the same time, that even if political themes are missing from the novels of the Western world, one finds many among the Vietnamese or the Palestinians. And it happens that all of the great novelists are on the right side in politics. With fine optimism, Carpentier affirms that, if there can be political engagement on the "wrong side," this is very rare. There have been, according to him, very few novelists, painters, and composers committed to the right politically (which is, of course, the wrong side). Nazi Germany is negligible, and we can simply slide over the Stalinists. However, it is really remarkable that politics is not the concern only for artists with a message but also for those who are abstract formalists, of whom we will speak in the next chapter. Philippe Sollers was a brilliant militant of the extreme left, even though, outwardly, overt propaganda does not appear in his writing.[3]

Painting, in like manner, uses a political discourse. One saw this during the big conflict about *Exposition 60-72*, which was intended to be a representative sample of painting from the last twelve years. Here it is evident that, compared to traditional painting, one juxtaposes objectivism and critical painting, both of which simply illustrate symbolically political and social discourse. Quite astonishingly these artists are outraged when the government tries to control such an exhibition. If one wants to be political, one should expect government intervention. Here, precisely, we can observe the role of this art that, indeed, reflects the technical system. So, what should art address? Politics in the narrowest sense. In reality, this art shows to what degree political discourse is the veil, the mystification, the ignorance with regard to the real. It takes on a passion that is a false passion, and, to the extent that it takes itself seriously by proclaiming a political message, we see to what degree it is impossible to take politics seriously. Of necessity, this is always the message from the left. But this art not only carries a political truth but also a militant truth that must engage the listener or the spectator in a certain course of action.[4] So, let fly all the trite terms of challenge. Art is worthy of the name only if it is a protest, a proclamation of total freedom (which is a theme directly spawned by technicism; *technique* allows for everything to be done, although it is here decked out in ideological-political dress: freedom against power, a return to 1789). The artists of 1970 paint Vietnam, the civil rights movement, the Third World, the protests of the young. Today, it's the environment and pollution that are in vogue. But, let's be careful: these themes conceal the growing power of *technique*; one attributes the problems to politics. If there's an environmental disaster, it is the establishment's fault. Whatever one challenges, and the most important matter is the challenge, one must make of it a political question in the name of the venerated but uncriticized principle, "All is politics." Crónica, the Spanish group, makes ironical political denunciation the essential part of its art. Of course, their challenges take the form of transgression or "sacrilege."

An art that does not embrace sacrilege would amount to nothing. What distinguishes above all the "immoral tales" from run-of-the-mill pornography is the over-riding, sacrilegious, and, hence, political intent. With transgression we witness a very significant and recurring phenomenon, quite characteristic, as we previously emphasized, of the old-fashioned nature of this art. Incapable of knowing today's reality, it transmits an illusory political interpretation of reality and, from that point, invites one to redo the revolutions of 1789 or 1917. Cézanne, Van Gogh, and Gauguin were aware of the transgression against the culture of their time. The early *poètes maudits* were also aware of it. Of course, one can say that Baudelaire was a *poète maudit* in relation to a certain milieu, a bourgeois and business milieu, but that he was a prince in the milieu of the Parisian intelligentsia. One can (and must) also say that the transgression carried out by Cézanne and the others was only the reflection, transcribed into works of art, of the much more absolute transgression made simultaneously by *technique* with regard to all beliefs, all structures, all traditions, and also to culture itself. Nevertheless, let us maintain the grandeur of these forerunners. Henceforth, it is taken for granted that there can no longer be art worthy of its name without transgression, without insult, and without aggression. Is there nothing left to transgress? Have moral and cultural taboos collapsed? Is the social body coming apart without defense? Quite the contrary. One will play at transgression. One will pretend to find the frightening dragons that guard the treasure of freedom in order to have the honor of combating them. One will loudly proclaim the odiousness of limits, of taboos, of interdictions, in order to rise as a hero of transgression. By in so doing, the artist bluffs his audience by raising the specter of monsters and shackles of another time, which he claims are current in order to play the transgressor. The important thing is to transgress, in whatever (*n'importe quoi*). Which leads us to an art that huffs and puffs and constantly changes both direction and style and which shoots blindly at illusory targets. Provocation accompanies, it goes without saying, transgression and challenge. One must involve the consumer, "in the dynamic process of the creative operation." Expressionism, above all, but afterwards all other forms of art as well are aggressive and provocative. On the one hand, the conservative or reactionary tendencies of the audience must be challenged and their guard must be lowered, but in the process they sink into ridicule and nonexistence. The audience becomes an object of derision and contempt and is called to abandon their untenable attitudes. And, then, on the other hand, the audience is incited to political engagement and participation. The audience is engaged by confusion, by the intolerable, by the absurd, by derision, by the attitude of anything goes, and by vulgarity. One makes a sculpture to provoke; one makes a film to provoke. Niki de Saint Phalle's film *Daddy* is a shining example, which is but one of a thousand works against the father. One provokes by attacking man in his most secret and guarded inner self, by attacking what he believed to be valuable, pure, and just. On the contrary, political denunciation lances the abscess and, evidently, makes it impossible to understand the affection of a father and daughter except under the guise of incest. If one does not spit in the face of the spectator, a production is not worth mentioning. But when one is goaded again and again in cinemas, theaters, and art galleries, one becomes impervious and also somewhat unhinged. In this amorphic situation of gradual immunization and anomie, one must keep increasing the provocations and transgressions. This so-called revolutionary art is, in reality, nothing more than the art of continual excess. It is true that art can no longer be the occasion for revolutionary protest. Thus, for many, the 1972 Exposition only justified itself as the locus for protest. This fits in with the idea of the happening that replaces all lasting and repeatable creative acts. But this idea of "always pushing further," an indefinite transgression that loses its meaning (*sens*) in a struggle against the absence of rules or against a fiction of taboos, leads inevitably to the transfer of meaning to nonsense, and to the abandonment of all possibility of a work of art and to dissolution. There is a direct link between the grouping of "challenge-transgress-protest, sacrilege, provocation," and the dissolution of all that was intended. It is a dissolution in the absence of what one is struggling against. The "young hot shots of painting" impugn, along with James Parmentier or Niele Toroni, every exhibition, every salon, which they see as a "useless show." In order to remain pure the work of art must be negative—for example, a hole in a photo or a reproduction, and therefore, it is non-art and non-literature that has the upper hand. And thus it is the obsession of the radical artist to deny everything.[5] To deny everything, not only society and government, but also one's self and ultimately the human. A total negation can lead only to suicide, if one takes it seriously, and if one does not, to comedy, for a denial of everything is equivalent to accepting all things. But, finally, whatever the judgment, we notice the tendency, which best expresses the fundamental confusion of all current artists (*cf.* Michèle Delaunay, *La ronde droite*, 1974) to capture the reality they wish to challenge. It is a mistake to think one can overthrow *technique* and its possibilities to challenge the established order. So it is with video as a "revolutionary" device and also with the computer. We are faced with a total confusion, a sense that one cannot know where to direct one's action: should it be against a society or culture handed down from above, an arbitrary centralization that rests on the perverse will of a few politicians, or should it be against "… the ideas of the ruling class," a struggle against the exploitation of man by man and against inherited culture. It is true that video can create official counter information and allows for the making of propaganda (Jerry Rubin in *DO IT* says this explicitly). It is true that this allows for an alternative to television. On the one hand, we have the simple reinforcement of the technical structure, and, on the other hand, we can sleep soundly. In order to destroy the structural elements cited above, one may simply let *technique* progress as it does. No need to get involved. These are merely part of the traditional elements of society, positive or negative, which are summoned to disappear simply through the play of *technique*. No need for revolutionary artists nor for art with a message. The technical system is in no danger from an art that is conditioned by that very system, an art of denial that propagates videos.

The work carries a message.[6] Yes, but we are in a society of nonsense, and the work of art must reflect nonsense. Is the revolutionary message nonsense, or, rather, is nonsense revolutionary? We are in a society where symbolization disappears and where the artist searches desperately for it. He can absolutely no longer speak a direct language, tell a story. He must be indirect. Nor can he any longer paint the real, because he knows that the real does not exist. He can only paint a refraction. But, what of the message? Is it clear? Is it accessible? We are in a society where everything has been put aside, with the sole exception of non-work, the snapshots of what has been abandoned. But, do we have a message? And if so, for whom? Is the non-work a revolutionary act *in-itself*? These are the questions that every contemporary artist runs up against. And here is a series of attempted responses, which oscillate around two great poles: on one side, art is conceived as the highest expression of society, and, consequently, to question art is to question the very society that produces it. It is unnecessary to have revolutionary art: the challenge is made in art itself; it is art that must be destroyed, and, in so doing, one dismantles society. An anti-art is, by itself, revolutionary because it reveals the true mystification. Therefore, the revolutionary artist is only able to produce an anti-art or a non-art, and, in so doing, he attacks society's essence. This interpretation ignores a single fact: that art is only a spectacle in technical society, and by attacking it, he simply undermines the spectacle. On the other side, art can no longer express anything but nonsense because technical society is totally deprived of sense or meaning.[7] When art no longer has any sense, does not represent anything, does not say anything, does not formulate anything, art becomes the truth of technical society. It is in this manner that the real character of technical society is revealed. Exasperating and decadent music is only the outcome of what is exasperating and decadent in our society. When we can no longer tolerate music, cinema, or modern theater, we may conclude that it is technical society that is intolerable. Thus, it is nonsense itself that is the message. Nonsense reveals the true sense or meaning of our world. Art at this moment fulfills its responsibility to demystify and to denounce.[8] This art in its entirety is a negation of the subject, of the author, of the creator (of consciousness) as the producer of meaning in the world. The artist carries that message more than anyone else, for, more than anyone else, he was a producer of meaning and a creative subject. But, if he is no longer able to assume this vocation, it is therefore the entire society that prohibits it and neuters any meaningful creation. This was, perhaps, one of the lessons of the great experiment of Bob Wilson in *Ka Mountain and Guardenia Terrace* at Shiraz. Nothing happens, nothing is understood. All is silence, concentration, and dance, a mishmash of scenes, images of forced symbolism—*prima facie* nonsense—with no explanation, in order to create meaning itself. Silence and babbling become a revolutionary act in a society obsessed with explanation and information, too full of discourse, but where meaning has simply disappeared. Once again, it is this silence and babbling which Tàpies wants to bring as a revolutionary message. "I have tried to attain silence directly … the symbolism of dust, of ash, of the earth, of solidarity that is born when one understands that the difference which separates us, one from another, is the same as that which separates two grains of sand."[9] When the artist appropriates the most ordinary things of life in his aesthetic silence in order to place before the public another message, he performs a revolutionary act: if he uses straw to create his "works," he is making the spectator understand that "there still exists throughout the world pallets of straw and that the artist conveys more interest there than to beds of gods, to their messengers or to the rich who worship them." The material itself must, therefore, vouch for the revolutionary choices of the artist. In a similar manner, theater, which can be pure silence, situates itself, at any rate if it is the bearer of a true message towards the absence of text. Max Frisch (author of *Biedermann*), who is called Brecht's heir, only believes in "theater without text" (going much further than *commedia dell'arte*): "The primordial problem is to allow the non-ruling class to create their own spectacle instead of imitating their masters as language would lead them to do … in order for the people to escape from the weight of their written heritage and from psychological naturalism it is necessary to resort to theater of the street or cellar, to mime experiments, and to participatory theater like that of Ranconi, Chaikin, Gregory. … Any longer, the theater can only serve to liberate the forces of revolt in the erotic arena or to remind one of the physical realities of life …." (interview by Frisch in *Le Monde*, 1971).

In the face of this Albee's attitude seems completely reactionary when he upholds the ideas of author, of work, of text, and when he continues to distinguish the roles of the actor, of the director, and of the writer. Certainly, "good plays" rebel against the status quo and convey anger, protest, and revolt against the order of things: therefore, this is a message, and a revolutionary one, because "all serious art aims to transform the world, the consciousness of the audience, and the world at large …" But, this is the message conveyed by the author. This is the position of yesterday's revolutionary artist. Today, we have the disappearance of the text (terrorist) and of the author (willy nilly the conveyer of the dominant ideology) that is the messenger of revolution. One does not question whether the directors or actors are also conveyers of bourgeois ideology: why just the author? But here we have the embryo of a response: it is language that is terrorism itself. Then the play is made of points of light, gestures, and contortions, of couplings and whimperings, etc; one is certain that, since language has disappeared, all the cultural baggage and all the contained ideological shackles … but when all is said and done these gestures, these explosions, these mimes are also still the conveyers of the same ideology. Making love on the stage is all well and good, but have not the erotic positions and techniques already been established by a bourgeois dispensation? One no longer questions whether "the people" (the true people, not those represented by the students and a few hauled-in union members) come to this theater appreciatively to attain a revolutionary consciousness. Be that as it may, if the ultimate goal of revolutionary message is the absolutely white canvas of painting, the total silence of the composer or the playwright, the raw material set out by the sculptor, an enormous amount of chatter proliferates around the creator's attempts.

For ten years we have witnessed an endless amount of artist's clatter about their intentions, their ideas, their proclamations. In other words, because the work of art, being unreadable, incomprehensible, ultimately non existent, the work of art requires an explanation to the public of what is happening and of what it all *signifies*. Indeed, with horror, painters reject painting that has a "subject." All know that one must ignore "the explanation," the "account." Novels avoid this like plays in the theater in order to recount an adventure having a beginning and an end—an interpretation appears as an aside. As in former times, there is a subtext: Bellerophon is carried off to heaven on a winged horse, etc. Now the subtext is no longer inscribed on the picture or on the painting.[10] If it were, it would be much too long. Now the artist's intention is set in discourse, in interviews, in articles, and even in books (Paul Klee, Antoni Tàpies) to explain those intentions. The work of art is so little a work of art (of course) that no one grasps it at first sight. I use straw because … If you see a zebra-like blackness, this signifies that … My theater simply stumbles about in order to express … Clearly, a long-winded explanation becomes necessary and shows that nonsense has become the sense in today's society. Concrete music is the world of sound that goes beyond us. Now, one must admit that at the theoretic level, that explanation created a certain interest. Then, in the presence of these "non-works of art," one discovers, with an intellectual labor, something of interest. Horror of horrors! This word has leapt from my pen, proof that I'm not detached from bourgeois ideology! It goes without saying that what is interesting is loathsome: art should commit you to revolutionary intervention and not arouse your interest! But, a turn to the author's explanation is frustrating. We encounter a "message" of radical weakness and lack of interest, which Adorno calls "mental debility." In reading nearly all the interviews in articles and books by authors explaining their message, we find such debility even in those who might appear the strongest—Godard, Beckett, or Robbe-Grillet. We have two major orientations. We have a hermetic accumulation of terms drawn from the social sciences as in Jacques Lacan or Jacques Derrida, or, we have left wing preaching that is side splitting, stultifying, and yet banal and does not go beyond the annoying "messages" that we find at the end of Chaplin's films from *The Dictator* on. These artists appear half-baked, incapable of an original thought, unaware of their own motives—spouting clichés with self-satisfied seriousness—unaware of their complete conformity to an already outdated ideology, and unaware of their complete lack of fit to the real world (B.P. Chéreau, Joris Ivans). I am speaking to you of the burden and struggle of class, of dominant ideology, of the overthrow of the father, of repression and sexual freedom; of *Les Nanas* by Niki de Saint Phalle that "portray the triumphant woman whose head is strangely small and whose other attributes are greatly exaggerated." Saint Phalle exhibits the essential truth that women are the last colonized population of the earth. Meanwhile, Tinguely talks seriously about the moment when computers take power. These formulaic reflections are quite shocking in their lack of seriousness, depth, and specificity. We are besieged by an endless, pompous blather, full of self-satisfaction from all these artists who boast of the depth of their works. This art only claims to be art by concealing its so-called message, because, in the presence of clarity, the message either disappears or is revealed as imbecilic. Consequently, I feel obliged to return to the work itself, which tries to say something but which produces nothing of importance. Thus, the novelty of the explicit "message" raises doubt about the validity of the work. But it is not only the political message that is ridiculous but also the spiritual or philosophical message.

The explanations of Stockhausen (*Le Monde,* July 1977) are such a tissue of perfectly hackneyed clichés that one wonders if such music should be heard. Lucifer at work in the factories, Saint Michael the archangel in the guise of a doctor or a good scientist, subatomic particles associated with the idea of Christ, the musical explanation of the Zodiac, the spiral of life and the history of consciousness, and finally, leaving earth for the cosmos, because our "island" is submerged in too many absurd problems. One is dumbfounded before the relationship between such musical science and such nullity of thought. If we consider the theater of the absurd or of ridicule, we find all the old chestnuts of a banal criticism: a bankruptcy of love and friendship, an absence of communication, a reign of terrorism, sadism, and the burlesque of *Ubu Roi*. How many times must one repeat that Jarry was astonishing, but the endless repetition of this message is simply ridiculous. And, again, Beckett, Arrabal, and Adamov have one thought, even if it is only the reflection of the technical world, but their followers are reduced to bumbling political appeals to an audience already won over to such noble causes.

We see the same incredible weakness in the majority of "happenings" when it is boasted that, in this conformist universe where individuals are isolated and passive, art should create spectacular events that seize the audience with respect. Such art creates events like these: a roustabout with a wig made of paper strips and half a dozen eggs that he deposits at the entrance inviting the audience to eat them or to break them; or someone asks strangers to take their hand and take their photograph; or a long line of foam is spread along a Parisian boulevard, etc.; all this is at the level of the student practical joke that one expects during snipe hunting week. All this shows the total and complete divorce between the incredibly vain, pompous, and "did you see me do it?" nature of the discourse and the tacky idiocy of this reality. These works are no longer the expression of art. They have neither form nor content. They hide behind a hermeticism that ultimately demands belief in a cardboard Easter egg. They are an aestheticism, which is outwardly rejected but inwardly embraced and which is completely mindless, and then artists put their message in a hermetic vocabulary, a second rate hermeticism, the better to hide the poverty of their thought. Put it all in understandable language (and I admit this is a form of terrorism) and you will see there is nothing. Take as the granddaddy of all this Henry Miller. You find in him all the clichés: the banal critique of society and its machines, its money, and its tedium; society is evil because it prevents the artist from living and expressing himself. Further, we have the high sounding ideology about art and artists, a portrait of the conflict between artist and Pharisee, and the notion of the ideal of the solitary artist misunderstood by the world, ending with the conviction that society will be transformed by art. But what a soft spot he has, what a bias for his United States, which permits one to find true nature and humanity. The entire American ideology parades through Miller's work: the ideology of man, of the south, of Sri Ramakrishna, and Swami Vivekananda; America is a great country thanks to its great men, its painters, and musicians. We find in Henry Miller a prodigious accumulation of clichés decked out by audacious pornography and by the false authenticity of the man who speaks his mind. And this man was the ancestor of an entire generation of writers, of poets, of painters. But what did he have to pass on? In other words, his expressions of revolutionary messages—his artistic expression of those messages—may be characterized by their incredible banality and by a pure propaganda understandable by all and accessible to "the people." This is pure propaganda that points people in a certain direction and convinces them of the truth and is characterized ultimately by a total lack of fit to the reality of our time. Miller's followers talk and think in terms of outdated analyses, valid for the historical *avant-dernière*.[11] Their message is neutered in advance by the very technical milieu in which it is forced to express itself and about which it is totally unaware (except at the most banal and superficial level, for example, the mechanization of man, the fear of the robot).

To the degree they are revolutionary *vis-à-vis* a false situation and false problems, their messages are either vitiated by the true power they do not attack (they attack the bourgeoisie, the capitalist, but not the technician, and they situate themselves at the political level and not at the primacy of *technique*) or they contribute to hiding the real situation. Therefore, the most exalted of these revolutionaries are in fact counter revolutionary. They attack bourgeois ideology, traditional morality, and the family … but this worked in 1880. Today, this is simply ridiculous. By so doing, they are concealing the conditioning process of the technical system in which they are shining examples of the enormous grind of *technique*. And when artists pretend to portray this in films like *Alphaville* and *Le Pain vert,* the content is so unbelievable that it is impossible to take it as a serious questioning of the real nature of our society.

But, one cannot neglect another pretension of this art. We see more and more the development in theater, in the novel, and in painting the art of the horrible—of bloodcurdling descriptions, situations, grimaces, colors, dismemberments, of inhumanity in all possible forms, of reduction of the human to a thing, of torture and abjection. We are saturated with horror. We have already encountered what is not presented as an excuse but which is, rather, a justification: "If the world in which we live is horrible, why haven't we realized it? Why wouldn't we make an art of joy—blue and pink, idyllic—when all around us is black and atrocious." The artist is a witness of his time. So be it. Let us say right off, that he is not, for this reason, a witness, for there are also admirable acts of devotion, joyful lives, happy loves, sane political efforts, living hopes. But our artists do not want to see those things. And when one of them does, he is despised as the conveyor of bourgeois romantic culture. Let us continue. This is not what matters to us here. In reality their overriding aim is: "By depicting the horror of the situation we push men to remedy it, we lead them to revolt, we provoke a salutary reaction. They see a dictator, and they throw themselves into combat against him. They see the unbreakable chains of money, and they are ready to join the fight against capitalism. They see torture and they will refuse to cooperate with any form of repression …" Such is the revolutionary aspect of representing the horror, which always has the face of the enemy that one must fight. Now, I believe that this entire thesis is fundamentally false and must be, to the contrary, unhappily reversed. In the context of the technological society, given the nature of man as we know him, the depiction of horror produces no positive reaction: it is simply an intensification of horror. Man experiences horror often unconsciously and at a low level. Suddenly he sees it projected in front of him, carried to a terrifying extreme, and he is told: "This is your fate." He experienced it as a kind of weight; he was shown its absolute destiny. He does not revolt. He despairs, becomes neurotic, and commits suicide. In order to soften the representation of the horrible, one needs a culture, a social or religious structure, for example, which transcends the situation and which converts the concentration of horror into a rite of purification. Purification is only possible with reference to something independent and greater, to something that transcends the situation. When the Middle Ages represent the eternal triumph of death, it is not horrible because death is placed against the Resurrection. One could look death in the face because it was conquered by the Resurrection. Now, however, the Revolution (with a capital R) cannot play this role because it sends man back to himself alone, and someone with a horror of death must recognize his radical powerlessness and inability to change the situation. *Technique* has conditioned man to be nothing but the plaything of objective powers and that is why they seem so frightful. All of this leads only to fear and to the desire to escape this "militant" spectacle. Man in the fourteenth century believed that death was defeated by the Resurrection; man in the twentieth century knows that the Revolution provides no final answer and must be enacted always and everywhere that it appears. There is a dangerous misinterpretation created by the artist's situation in relation to what art signifies. The representation of artistic expression is not a rite of purification, of exorcism or catharsis. Man is affirmed only by his impotence and by his continual domination by evil. This art no longer attempts transcendence but is, rather, an imprisonment, the counterpart to transcendence. Transcend what? What can be transcended if one makes, *thanks to the perfection of technical means in art*, a representation that is absolutely immanent? Where would an immanent transcendence go? From the painter of horror we have nothing but a horror offering no way out. It's a cry in the wilderness. Only the blind can see. Man can only give in. Performing *El Presidente* in a theater can only lead the spectator to abjectly accept dictatorship, and the more monstrous, absurd, and incredible the performance, the more the viewer becomes small, invisible, and untouchable. Not existing is the only solution, the only fall back, in a world dominated by atrocity, its domination so complete that art can only talk about it. Such talk, shaped by *technique*, takes only temporary refuge in eroticism as an escape from *technique*, but this only doubles the simultaneous horror experienced through this art. Man finds himself bereft of anything that might, at his humble level, allow for a plausible life.

III / *The Communal Spirit*

Art with a message is still, when all is said and done, propaganda. And one of the most interesting aspects of this propaganda is the communal aspect. One must not have an author or spectator, a producer or consumer. One must suppress the distance between the consumer and the work. This art must be a coming together, a communion, a collective creation. People dream about Greek theater, which they greatly romanticize. But it was not the theater that produced the community: the community existed in the city and the drama strongly expressed community sentiment. The painter provides a blank space where the public can express itself. The attendees (I do not call them spectators) at the *Cartoucherie de Vincennes* are called to directly participate in the revolutionary parades and songs. Meanwhile, others receive from the "actors" touchings, caresses, solicitations. And in the Living Theater they pass around a loaf of bread symbolizing the communion of all. The poet hands out texts to be reworked according to the rules of the game of Go, or, alternatively, lines of poetry that you can combine at will (like the one million poems of Queneau). Even architecture should cease being a ready-made product that the architect delivers to people in its completed form. One must have the possibility of modifying one's apartment by changing the partitions, etc. and by arranging the space to one's own tastes. Or, to go further, Ricardo Bofill, with his *Taller de Arquitectura* in Barcelona, seeks to make "the people" participants in drawing up all the plans of the large buildings that he intends to construct. But we do not know precisely who these people are nor how the consultation is carried out. Simply *afterwards*, we have seen that they are asked for their opinions and evaluations on the possibilities of living in these buildings. Communion. Participation.

One of the essential concerns of the community is to make the public and the consumer participants in their own right. There should no longer be a completed work but simply a beginning to pursue, which would be clearly exemplified in the theater where one had spectators participate in the performance of a text that did not exist. Hence the Cell for the Creation of Open Theater (1978), where spectators created the play by giving their opinions and making suggestions. Of course, this goes as well for all the arts intent on providing an audience for the uncompleted work: a brushstroke, a piece of roughhewn material, and they become either a painting or a sculpture. It is the spectator who must make the effort to provide all that is missing … A triumph. No more passive spectator!

This is the creative activity of those who were once passive bystanders, consumers, or listeners. The technical devices are also revealed as wonderful possibilities for creating. From now on, the artist is only an inspirer, a game leader. "Nineteenth-century art demands an empathy … hence, an active and creative participation from the spectator … a more temporal awareness of being and acting: the values of the possible rather than of arrested, normative being … a pictorial field open to floating structures that can go beyond expected categories." (Delevoy). For instance, Constantin Brancusi and Claes Oldenburg, both, call for the participation of the spectator. The spectator, seeing his reflection in the polished surface of a Brancusi sculpture, seems to penetrate its form. And sometimes the work of art, in order to appear, needs the spectator to pull levers or to step onto a moving platform. In the theater (because traditional social relations have lost their value and humanity) authentic, new, social relations are created between spectators and actors. The new theater is a theater of participation, establishing above all communion. It's no longer the responsibility of the theater to make one think, as in the time of Sartre and Camus. Nor is it necessary to make the audience feel something, or to give a message, however indirect, through a series of emotions or through a series of images, however incomprehensible, as in the work of Beckett and Buñuel. For this to *be*, the distance between the spectacle and the spectator must be suppressed. In the struggle against the society of spectacle, one no longer practices a distancing from the theme, dear to Brecht, but, on the contrary, one submerges the participants, actors and spectators, their roles having been dissolved to become agents who exalt, who meditate, and who commune in reciprocity. The Living Theater attacks the spectator with a mock counter attack against the advertising of the mass media. "One turns the spectator himself into an object so that he can enter the play as a subject."[12]

It is true that one can mean almost anything under the rubric of public creative participation that involves no separation between artist and audience. Have we not heard one of the directors of Beaubourg declare on television in September, 1976 that acts of creation and participation were accomplished merely by a stroll through the galleries. A public parade is already, in itself, an artistic event and an act of *participation* by all is a creating of a work of art. This is clear as day, but not without a second thought.

Theater seeks to create a communion that is life itself, which is carried to an extreme in the "rituals of the counter culture" or the underground. The communal factor is particularly intense here, albeit a manufactured communion created by technical procedures: speed, hallucination, cross-dressing, earsplitting amplified electronic music, strobe lights … In particular, pop music has claimed to bring a communion, and certainly the concerts on the Isle of Wight, at Woodstock, and others, suggest this. Communion in a solitude of others, however. Pop music has been a music of communion and, one could say, this music has become a springboard for transforming morality, social life, political and religious thought.[13] It took up the cause around 1953 with Elvis Presley's "white Rock and Roll." It was the music of young people on the margins. They were a community who opposed adult society, a consumer society. Music was the means of affirming their marginality. It created a mystique, and one could suppose that the great anti-technical mutation or change had taken place. This music captivates with piercing sounds and possesses a unique penetrating power. However, this music is essentially without perspective, a kind of plasma bath that seems to arise from an incoherent spontaneity, which is, ironically, totally integrating because all the participants experience simultaneously the same incoherent spontaneity. It seems to them that it is their own expression because everyone is vibrating in tune or resonating to the tuning fork of instinct and intuition, and the event is created through the improvisation of all. A new life style was discovered.

Without doubt one could quibble about the success that produced millionaires like the Beatles, the Rolling Stones, and Bob Dylan. But this quibble is insignificant when compared to the authenticity and spontaneity springing from an astonishing conjunction between a message profoundly rooted in reality, a music generating ecstasy, mystique and communion, and a youth movement ready to kick the traces to enter a free spiritual life. An art immediately accessible to the people was opposed to highbrow culture and art. It expresses the hopes, the disappointments, the doubts about life, and offers reflections of the great upheavals of our time. The result is a pursuit of absolute freedom (freedom, freedom, as Richie Havens and so many others sing), in songs about Vietnam and against all wars. A quest for an absolute. "Nothing but an Absolute." One can't deny or minimize this great outburst or claim that it was a flash in the pan. There has been much talk about what followed this outburst. The pop movement became good business. Show business and record producers took over this enthusiasm. The avalanche of advertising propagated the mass gatherings and generated the hit parades. The idols of the recording studios follow in the wake of the movie star system. The love fests became concerts attracting select fans and producing elevated prices. The underground formed part of the normal landscape for the average man who read *Paris Match* and who watched television. By 1970, the affair had gone bust. But it is not the commercialization or the trivialization that concerns us. Realistically, this process was predictable without the intervention of capitalism. In pop music we have certain characteristics such as binary rhythms and electronic instruments. The electrification of the guitar and then of the piano are decisive factors that modify the acoustic sensibilities of the listener and alter the stylistic propensities of the musicians. "The electric guitar has called for short modulations on portions of tone," while the overwhelming power of the music removes all individuality and reflexivity (L. Malson). In other words, pop music is first and foremost a product of *technique*. We come face to face with the remarkable phenomenon of a longing for escape that attempts a radical critique but which, nonetheless, links up with the technical system to which it owes its very existence. Here this medium is a hundred thousand times more important than its message. The sense of communion, even if real, is merely the expression of technological society. This is even more evident in pop painting, which, in its intercourse with the products of industrialization, utilizes all the elements available from the most modern techniques in its attempts to contest technical society. It's the same song and dance that one invokes to produce a participative culture out of a process like videography. One seeks to eliminate the passive spectators to reach the living participants who will create in the form of communion a community. The use of video recordings, which can be erased and used by anyone, is bound to modify relations and to lead to a rediscovering of a co-humanity. A reality that is new and relevant, then, must be presented. Remember, though, that the focus will change: there's no longer any interest in a totally transitory work nor in its purpose or aim, but only in the *process of production,* which, it itself, becomes a thing to be experienced and not a product. In other words, this process of production or technique itself becomes experience like the "dripping" of paint from a brush not touching the canvas, as in Pollock's work. The process that involves physical action allows an authentic encounter "produced during the time the experimenters live out this utopia in their acts of creation (*experiment-action*); the work will be displayed to the people who offer their interpretations of it, with an eye to push further into the utopia and its embodiment (interpret-action)." (Willener, Milliard). The process demonstrates what happens with pop: *technique* producing communion producing popular culture. Thus this entire direction attempts to be communal by allowing the rediscovery of human authenticity and by claiming to be a protest against the technical world, but the result is nothing but *technique* itself; this art is determined and defined by *technique*. One must realize that *technique* is the producer of a communal sense. Needless to say, a sense of community is created between the practitioners of the same *technique*—for example in the fraternity of first generation airline pilots and biker fraternities. *Technique* creates a society of spectacle that is communal in some depth. Debord is particularly good at showing and analyzing the process of integration into the spectacle and into communal sense without true communication that is characteristic of the technical process producing an "absolute power inside a system of language without reciprocity," and one can also recall McLuhan's Global Village, which was entirely founded thanks to television. Individuals are pulled in because television is a "cool medium." The television viewer is thought to be isolated and solitary, enthralled by the small screen, and in a way this is true. He has no relation physically to those by his side. But there is communion with millions of others through an in-depth participation: television engages people with each other more profoundly than ever. "The industrial giants of America have had to redefine their goals and images to increase the engagement of their public." "Our ten-year-olds, when they sit down next to MAD, show us that the television image has pushed American culture beyond its consumption phase." Television produces by itself an image of the mad and ridiculous world, but at the same time, regardless of the product (a political message, an artistic work, an advertisement for a commercial product) produces a fascination and desire more important than the thing desired as the participation of the public grows. Audience participation is what counts. Television seizes you. "It grabs you. You must be in the know." (Hence the term "in" with reference to television). One knows that McLuhan bases his analysis on the necessary involvement of the spectator in television.

The spectator is bombarded with impulses of light. The television image is content poor. It is not a fixed image but a contour endlessly formed by the cathode tube. The image is composed of three million points per second, a mosaic impossible to grasp as it is. The spectator registers a few dozen of these in order to form a picture. He makes a kind of unconscious choice for a continually changing image. The television image makes us "fill in" the blank spaces in the pattern. "Young people who have undergone ten years of television have contracted an urgent habit of in-depth participation, which makes the long-term goals and visions of current culture seem unreal, devoid of sense, and anemic. The mosaic of television teaches the youth a total participation in a totalizing "now," outside of which nothing exists. The change of attitude does not depend in any way on the content of the programs … The child of television dreams of participation and does not seek a specialized future profession. He seeks a role, an in-depth engagement with society …" (McLuhan). Whatever the criticisms directed at McLuhan's theories and his ideology of the "Global Village" fostered by direct communication, he has clearly seen the participatory character of television, which is situated in the communal thrust of *technique*. Thus, not only are we able to say that the communal spirit in modern art is only possible thanks to technical means (which we saw previously), but, furthermore, it is the technical system, which, under diverse forms, implies and engages communal forms of relationship. There is a mechanical production of communal spirit.[14] Thus, art which sees itself as the bearer of communion is only a simple reflection of the productions of the technical system and which, furthermore, it needs in order to function and to be tolerable. This is why Charbonneau can demonstrate the point at which communion and participation in this art are mendacious. "Traditional cultures were at first participatory while ours seeks enjoyment and pleasure. Whether this enjoyment or pleasure is aesthetic or grossly sensual is secondary … Our organized culture is constantly threatened by monotony. It is cut off from the inexhaustible flow of nature. By multiplying aesthetic shocks that cancel each other, the slow germination of forms that express the life of man is prevented. Of the innumerable societies with competing styles that multiply, there now remains only one: this is not the eclecticism that simultaneously collects and refuses the art of dead societies, which could replace the spontaneous syntheses of technical society." But that is precisely what communal spontaneity claims to do. It claims to restore the possibility of a new outpouring and the reclamation of society by means of aesthetics. Street festivals, a breathing of new life into culture, are movements that allow people to tolerate a technical society that is no longer real and to compensate the frustrations of its mathematically rational solitude and enslavements. But art remains, serving its same function, at the same time paying tribute to society from which it also separates. It is, as it was in the time of bourgeois luxury, the supplement to the soul, which is sold dearly.

IV / *Compensatory Ludism*

Play finally enters into compensatory mechanisms produced by the system.[15] Art becomes a game, an invitation to play, at the festival. One now denounces the most lamentable lack of festivity in our society. This ludism, which is justified apparently on the basis of Johan Huizinga's *Homo Ludens*, has the same nature as the communal tendency wherein one finds the return to intuition, spontaneity, and pure irrationality. Assuredly, the happening is part of this ludism as a protest against the rational, the anticipated, the calculated aspects of modern life. It takes the form of entering unexpected situations contrary to the acquired habit of always placing ourselves in the familiar where we already know what actions are required. The happening is an impossible and ridiculous situation where one is taken unawares, and each one is obliged to create his own response. The happening explodes one's defenses and one's cautions. It creates anguish and insecurity, which leads to reflection. Likewise, we witness a new romanticism of the imaginary, of the useless in art, of sentimentalism, of intuition. Louis Kahn, one of the greatest American architects of the 1960s, seeks to start with a clean slate, to pretend that there had never been a school or a hospital, to approach everything with a new eye, and that can only be done through intuition. The same goes for Roger Bissière in painting: "The hand advances into the unknown … nothing can prepare the canvas. We are in the realm of pure spontaneity." And, consequentially, nothing can explain the picture! The painting is beyond explanation. There is no concern for description. One can suppose this painting is an act of play, and we rediscover that it is a tragic act of play. One of the great aims of play is precisely the refusal of the rational, of the organized, of the planned, of *technique* itself. Many artists will repeat the slogan of one of the Bauhaus masters: "Where the machine goes in, man goes out." The architect rejects the construction of cities to accommodate the automobile. We must restore open areas, space to breathe and to play. Play in architecture is one of the more important elements that Ricardo Bofill reaffirms, and Pop art was also an act of play. This art needs neither content nor meaning. It is meant to compensate for the frustrations imposed by contemporary civilization and society. Everything that is repressed in other ways is expressed metaphorically in art. Since contemporary culture is repressive, art must provide a counterculture of freedom, and this is found at its best in the act of play that reproduces raw experience and engages in provocative paradoxes. This, then, leads us into an "informalism," and, to a certain degree, Op art itself, despite its rigor, transmits no information, describes nothing, and is content to produce mere sensations. Forms are fleeting and are arranged through play; words clash in uncommon ways; painting becomes pure gesture; one must capture the random, the unrecoverable. One must play like young animals and produce art while playing in this way. And so, if one plays, one understands that there is nothing to understand. Hence, to let one's self go and pursue one's pleasure and one's desire. Since, in our society, there is no longer play or festival, art is called upon to restore this human dimension. But, what is more, ludism and festival are revolutionary affirmations, and, in this sense, they challenge society. For our society is totally geared for the serious, for the efficient, for the timely, and to introduce the unexpected elements of festival and of art is to bring all into question. At this point we find the rather tragic character of art that seeks to be a game. It is then over extended and the proposed festival is excessive. In modern art you are invited to play, to go to the game and the festival, with an evil eye, an insult on the tongue, a contemptuous face, and an exaggerated tone. If you do not join the game, you are an ignoble, reactionary bourgeois. Play is a serious act. When Robbe-Grillet proposed his *Project for a Revolution in New York*, it was a game where the cards were intentionally scrambled and where the rules were not known in advance. This could be applied to dozens of modern novels. And, of course, Niki de Saint Phalle and Tinguely considered their work in the category of play: machine-sculptures animated by mechanical jerks … or gigantic dolls, exuberant toys. In this game the artist plays and invites the general public to this game meant for the public. As part of the *Festival d'Avignon*, which has become an immense fair, a people's aesthetic parade is held to express "freedom." Other festivals like that of Avignon, since 1975, are typically a quest for a communal spirit, which is tragically transitory, and arises from the illusion of staging a revolt, of mounting a radical protest—who knows against what? Apart from being fashionable, these events are important because they reveal an intense desire for change and away from despair, all of which is fundamentally ideological and hardly viable. These events conclude with a milling about of bodies set in motion by a willful blindness to their futility. First we have the Game at Liberty Square and then the video of it. With the video, you can do anything you want and express whatever you want (*n'importe quoi*). But the reality of this play provides the validation of man through the technical device. Let us remember the effect of video on those who become models and actors: their play is motivated by the subsequent projection on a screen that makes them important, people of interest, stars. Here freedom comes from the machine, but as in all cases of this type, this freedom exacts the price of a hidden servitude. To the degree this play is free, it engages us involuntarily in an aestheticism. The American novel has fallen into aestheticism through its pure and simple will to play with no intent of social reform. Truman Capote, Salinger, Updike are significant in this regard, but we will deal with aestheticism in the next chapter. At this point, note the link that unites play and aestheticism at a certain stage of technical evolution. Ultimately play puts us in a setting beyond real life and ludism adds the unexpected to this scenario. This explains the painting and decoration of walls, billboards, and utility poles. This mixture of setting and play can be found on the walls of Paris painted by well known artists, such as César, but also with the authorized participation of children, students, and fresco fans. For example, on the length of a billboard in *Les Halles* we see flowers, birds, butterflies, which have been drawn by primary school children. One also sees auto parts attached by blow torch and turned into elements of the setting … But all this, which is definitely a form of artistic expression (and what isn't nowadays?) and which expresses a play of the people, does not get beyond mere setting. A happening can last two hours, and it is certainly not bad to have two hours of play. But this does not set aside reality and is merely a timeout or a compensation. When we bring up the question of play, one cannot forget that in the nineteenth century art was viewed in this way by the bourgeoisie. Art was to be an aesthetic appendage or an agreeable addition to utility like the metal flowered sunshade on McCormick's reaper. But what else are the painted or decorated Parisian walls? What is new with respect to this absolutely dichotomous point of view? These decorations compensate the difficulty of living in an unforgiving reality. The art nouveau of 1900 has already done this. It has been criticized for "having turned the useful into ornament and having tricked out the useful with whimsy," but isn't this what the play of modern art proposes? One could say that art nouveau was a "parenthesis dedicated to aestheticism." This is wrong. Aestheticism was the compensatory reaction *vis-à-vis* the development of industrial society, which was found intolerable. In this same way, the painted walls are a protest against dullness, anonymity, sameness; they are the glimmering of a false reality that forgets the real. Michel Butor's *Réseau aérien*, a work written for radio, is typical of this compensatory play against reality. It demands a detailed reconstruction of an appearance of life (following a couple in detail during their trip by plane to Nouméa, with all the complexities of life itself), by totally artificial means, directed toward listeners who will never experience what they are given to understand in an admirably realistic fashion. Once again, art allows vicarious experience through a third party and thus becomes an enormous bluff. Above all, the real is hidden in a showing of the real as an illusory and reconstructed reality, again for one's amusement. But, during play period the student forgets the classroom, which is the point of compensatory ludism. We must then become conscious of the two sides of ludism. First, it is a *compensation* and therefore exists only in relation to *technique* itself. Play is defined by the orientation and impact of *technique* and varies with the frustrations imposed by *technique*. It does not, therefore, allow a transformation of society despite claims to the contrary. Play challenges nothing. It's an agreement with an escape to that which cannot be reclaimed. One believes that one has won because, by bringing art back to play, one is given a false refuge. It is not the invention of a new possibility; play is a refusal of the only possible reality. It is not a real critique, because play activity can be tolerated and absorbed indefinitely without any change to the system. The reactions to this play are only skin deep. They are the grumblings of people on the sidelines … and that's all. Play on, play on: play the flute, play at painting, play at theater, and all this time the established order remains. Those who play think they are free in playtime, and to the degree that people experience this need to play (and we are not talking about playing the stock market), they fulfill a useful function in a society devoid of play. The players can be tolerated and even honored. The great players like Tinguely, César, or Mathieu, are they not honored by the most technical of governments? Technocrats are convinced that aesthetic ludism is indispensible, and they are ready to back it. Meanwhile, this play, broadcast by the mass media, gives the spectator an intense vicarious feeling of freedom flowing throughout the society: one can do anything, even a sit-in against the Concorde or painting the walls. But there is another side to this play: ludic art is increasingly separated from reality. To rejoin "life" and to escape the machine, ludic art, simply, abandons social reality, while claiming, as we have seen, to remain revolutionary, a claim which is nothing but redundant and wordy and only reveals its impotence to confront effective reality. A fresco about American imperialism in Vietnam can be painted. But the profound reality of the world in 1975 reveals the growth of *technique*, which art ignores. Ludism shows, albeit unintentionally, that art, which claims to be politically involved, instead, plays with a politics of "elsewhere and yesterday," rather than one of the "here and now." Involuntarily ludic art points to its incredible vanity and its gratuitous and vacuous nature, a witness to its own uselessness. The game is played in opposition to the reality it cannot change. Consciously or not, the game is simply escape, and modern art does not go beyond that. It is exhausted in demonstrations that are only entertaining and distracting, while claiming to be absolutely revolutionary: an art of the Molotov cocktail that is nothing more than a child's firecracker that is a pretense of war. Technical society already proposes serious games (urban games) or non-serious gains like those of the mass media. Technical society is simply beside itself and gleeful when the artist, the creator, the innovator enters the ring to entertain the "stupefied masses" and to inject them with a new lease on life and a new interest. This kind of society needs aesthetic ludism and a cultural enthusiasm. Is it at risk of destruction? What an illusion! Take for example Pop art, which was a recent expression of a popular revolutionary ludism. What were its principles? The organizers of the grand exhibition of Pop art in 1969 summarized them thus:[16] the overthrow of all pictorial conventions (but what is challenged? Painting and nothing more!). It is the substitution of industrial techniques for traditional oil painting (hence the primacy of *technique* that aligns art with its most advanced processes.) All the hierarchies are leveled relative to the subject and, therefore, according to the bias of the technical system where neither substance nor creator has any importance, and it is not the artist as subject who counts. Conformity to the technical system is uncontested where one does not ask what one does; one simply *does* as well as possible. Anyone having the right technique can perform as well as anyone else. The subject disappears, not for philosophical reasons, but because the technical system eliminates the artist from life as such—exclusion, elimination from the world of dreams. This seems to contradict what we have said of ludism: there is an apparent reassertion of reality, a denial of surrealism, but it is an assertion of a false reality. One plays. But one plays with the rejects of the technical process. Recusing the dream is not refusing flight; rather it's practicing a more dangerous flight because one no longer even realizes that there is flight. One believes one is taken up by reality. Ultimately, we have a much more *noble* concept of art.[17] But, isn't that just one aspect of *technique*? Thus, by formulating these principles, the theoreticians of Pop claim to combat the current situation, while, in reality, they are its reflection, and, furthermore, by claiming to reintroduce play, they only engage in a diversion and an obliteration of the real under the guise of a totally free imagination and critique. Under the guise of reaffirming the person and of expressing the people, they engage in vast protest. The sterility of art with a message comes meekly to die on the shores of the unforgiving beach of an indefinite system.[18]

Notes

1 There was, in an article in *Le Monde,* October 1971, an excellent parallel drawn between two concepts of popular theater in 1950 and 1970 that showed perfectly the evolution of art concepts in general but also the impasse in which art is engaged and from which it tries to escape through an excess that permits an uncompromised return to a reassuring and rational technical universe after the aesthetic madness: a fine technique, a useful animal, a comfortable jalopy, the joy that things are as good as they are. Damn, we have just left a nightmare … It was only theater, or painting, or music … a real good movie. Here are some comparisons we'd like to draw:

From One Dream to the Other

The two generations of popular theater are more alike than they would like to be. Could this be in their common origin of "defrocking the bourgeoisie?" The future will certainly bring them closer together as a result of inevitable revivals and reevaluations.

But, it is a fact that the new theater is defined in large part by its opposition to the old to the degree that one can list in two columns what theoretically divides them:

The Fifties	The Seventies
	Or
New auditoriums	The street and workplaces
Without a proscenium arch in the Italian style	Without any fixed structure
Directed toward the audience	In the middle of the audience
	For Whom
The greatest number	Small groups
To attract the audience	To go toward the audience
A mixture of all classes	Primarily the working classes
To shape the spectator	To submit to the spectator
To build an audience	To build instantaneous connection
	By Whom
Artists to the rescue of the public	Artists in public service
Legitimate heirs	Willing bastards
Artisans	Amateurs
Specialists in expression	Spontaneous militants
Directors	Agitators
	For What Reason
Culture for all	Political combat
Cultural democracy	Permanent revolution
Better order	Disorder
	How
Good theater in itself	Theater in itself does not exist
The classics honored	Classics to the scaffold
Eliminate bourgeois elements	Eliminate all heritage

Exalt what unites	Show what should divide
Seek the universal	Denounce class bias
Lyrical illusion	Realistic message
Reconcile the hero and the world	Invitation to combat
Change relation between play and stage	Change relation between play and audience
Passive appreciation	Active participation
Cult of the hero	Exalt the group
Ritua	Festival
Moral education	Physical outburst
Rigor	Exuberance
Necessary utopia	Possible triumph
Abundant theory	Violent practice
Applause	Slogans
Reflection	Action

VILAR: "Create a good society after which we will perhaps create good theater."

BENEDETTO: "Perform theater with the aim of creating a society in which everyone will create their own theater."

2 An excellent and ironic little note by Paul Fabra (*Le Monde*, November, 1965) emphasized that, quite surprisingly, literature is almost uniformly pessimistic, whereas the works of economists are remarkably optimistic. He contrasts, among other things, Orwell's *1984* with the report of the "Horizon 1985" group or with the well-known Sommers report. I could add to this the bulk of sociologists.

3 Before the palinode of Sollers, perfectly explicable, moreover, its function remains the same.

4 It must be said that in art with a message, there are two criteria of judgment: in countries where democratic or republican liberalism hold sway we see revolutionary messages shouted explicitly in a deafening roar and proclaiming the urgency for tearing all down. But the artist can be prudent; in dictatorial regimes, whether left or right, the message is abstract, symbolic, and often undetectable. Thus, Tàpies will take impoverished materials and colors to symbolize the literal impoverishment of the people oppressed by Franco's dictatorship. He will paint a red and black picture to denounce the blood and death produced by the dictatorship, and so forth. But this symbolic language is so obscure that no one, neither the people nor the police, see anything in it. The message has only liberated the consciousness of the painter. Prudence is the mother of safety. It is the watchword of modern revolutionary artists with a message.

5 The act of art par excellence will, thus, be to dump a pot of red paint on the artist (Malraux) and then proclaim this act in photos at Beaubourg (February 1977).

6 Those that carry a true message are those who do not claim to. The admirable André knows what he is about in his "automatic" designs. There is no need for theory or explication: *he simply is*. He is so deeply rooted in underground art and philosophy that he has no need to beat his brains to employ great words of political-revolutionary messages: his painting speaks for itself.

7 When Tinguely utilizes mechanical elements retrieved from old jalopies and other things to construct his own machines, machines that clearly signify nothing, he can claim to reveal the absurdity of the industrial world, but it is the industrial world that has its revenge on him, for it is he who is absurd. Or, rather, he can claim to introduce poetry into the utilitarianism of machines: in which case, he does nothing more that what the bourgeoisie has always understood about art. Art is the gratuitous poetic supplement, which one can do without and which is subordinated to Technique.

8 So that there will be no confusion I specify that I am reporting here the discourse of revolutionary artists and not my own. I believe, to the contrary, that we have a case of false revelation, which is only the mechanism of compensation and a mystifying demystification.

9 See the remarkable *La pratique de l'art* (1974). Here Tàpies probably is the ideal expression of an obsession for the material, the suggestive power of matter and its force of simulacra—worn out wood, tortured linens, brown earth, tar twisted straw … "Tàpies has undertaken to 'paint' with matter rather than with colors, materials which have quite naturally become objects, things that are concrete and real. A work by Tàpies combined with his ladders—real or simulated—his trestles, his boilers for clothes, his hats, his big Xs, his outlines of bodies, amounts to a setting up of things and their surrounding space that actors—present and absent at the same time—have apparently deserted." (J. Michel). But, this lyricism of matter is, in its apparent innovation, the nostalgic backward glance of the artist for a dead past. Matter no longer exists in its specificity due to the ascendency of technique, and it must be preserved, ennobled, and displayed, henceforth, in the work of art as the unconscious witness of a lost paradise.

10 Nevertheless, we see how titles of works by great abstract artists represent a discourse: *Still Life in Motion; The Chirping Machine* (Klee); *Achilles' Skeleton* (Arman), *The Fate of Animals* (Marc); *The King and Queen Surrounded by Running Nudes* (Duchamps) and the innumerable and fascinating titles of Picasso's work … Therefore, there is a "subject!" (in the sense of a painting that has a subject).

11 I have written at length in my two studies on revolution, but, here, clearly, I am aiming at artists, intellectuals, and revolutionaries of the Western World, Europe and the United States. When painters, novelists, poets, playwrights, filmmakers from Latin America attack in their works bourgeois domination and American imperialism, they are clearly on the right track and their message is truly revolutionary *vis-à-vis the situation in which they find themselves,* but the error and the lie begin when those works are repositioned in Europe in order

to convince us that here, also, this is really the basic revolutionary problem. Jacques Ellul, *Autopsy of Revolution*, 1969, and *De la Révolution aux Révoltes*, 1972.

12 On the contrary, numerous initiatives "meant to put the people in charge" are entirely positive, inasmuch as they avoid the machine or the motivational specialists who run everything as at Beaubourg. But, there are, indeed, authentic experiments which are largely unknown. Thus, the worker theater of Montbéliard or the celebration of Mardi Gras and Midsummer Night assume the character of a new type of a non typical people's celebration with well wrought theatrical works that are neither spontaneous nor textually based. The gestures and double entendres are immediately understood by an audience completely complicit with the authors (who are also not professional).

13 *cf.* for example, Albert Raisner, *L'aventure Pop*, 1974.

14 Of course the relation to society is in reality both involuntary and unconscious. Very illuminating on this subject was an investigation in *Le Monde* (July 1975) on what the real and realism meant to television producers. One is astounded by the intellectual pretension, by an ignorance of true reality, and by an inability to get beyond the traditional debate about realism manifested by these producers of television. All was dissimulated under a vast, inconsistent, pseudo-philosophical discourse. Not one of them seemed to understand technical reality even in their own field of television.

15 Let us indicate at once that Ludism has two very different aspects in modern art. One is compensatory Ludism, a reaction against the system. The other, which we will see in the next chapter, is gratuitous Ludism, an expression of the system.

16 *cf.* The catalogue of the Pop exhibition at the Hayward Gallery, *Le Monde* (August 1969). Here, the message of Pop art analyzed as a reaction against the dehumanization of the world, but caused by what? Is it against mass culture, against the worship of material possessions, or an effort to revalidate self and being? How quaint. This superficiality does not go beyond a lyrical sentimentality or a romantic revolt … expressed through an insertion of technique! H. Skoff Torgue, *La pop-music,* 1975.

17 One must not forget that in order achieve engagement, art is as ephemeral as the event that inspired it. It can neither discern the essential element to attack in this society nor find a critical distance from it (in spite of theory, and we can't forget that Brecht failed to do this with Stalinism). Without critical distance toward the components of this world technical art is incapable of symbolizing or lifting anything to a universal plane; it simply dissipates without a trace. One can no longer characterize this dissolution brought on by its own deficiency (and not by the absence of state support, which these revolutionaries expect). The Festival of Avignon of 1976, with its ridiculous choice of street performances, is a good example of "whateverism" that pretends to be engaged but is occupied in a screeching insignificance, in blather, in gesturing, in a false celebration, in true incoherence, but a real moneymaker killing committed theater.

18 These pages were written well before the appearance of punk and disco. But these movements are nothing more than the conformation and continuation of the previous movement of insignificance congealed by the hypnotic effect of technique. The sounds, the shouts, the gesticulations, the frenzied outbursts, the throbbings, the fragmentations are in reality perfectly stereotyped and express a programmed type of music. The sounds that burst forth do not express any "emotion" in spite of what one says; they simply produce an instant of mind-altered happiness. One must not forget that after punk the emotionless style prevails. After the anarchy and spontaneity of punk comes, not with a tip of the scale but with the continuation of the same tendency, a frozen, rigid, and petrified style. It is not for nothing that one hears "Do the Mussolini," a derisive appeal to Fascism, a constant appeal to death, "I wish I could die …." What causes this completely depersonalized and neutralized ethos and a glorification of militarism? This is not by accident. Death and war are prefigured in these frozen aesthetic modes, trapped inside the ice cube of technique. Another example of the art of horror is a singing group called "Suicide," a synthesizer and a voice in the dark. This is music that conveys the absurdity of nothingness, the power of incommunicability and absence in a voice without passion. The singer is uninvolved and does not follow the lead of the synthesizer—a composition with no beginning or end that suggests the extreme limits of human absence.

Chapter 4
Formalism and Theory

The other major direction of modern art, opposed to what we have already described while still dependent on the technical system, is characterized by formalism and theory. This art without content, without message, without meaning sometimes mimics *technique* either directly or indirectly to produce a hermetic and academic type of art that plays at play but is merely a play for specialists. It is no longer play for the people or for children drawing on walls or setting off fireworks. It is like a game of Go or chess. It is an art that is more and more refined, following the technical model; more and more difficult and strange to the unschooled populace, but it is thrust upon the people who only understand it superficially without appreciating its amazing subtlety. Only the specialist can interpret it correctly. This presentation to the public is of least interest to the artist: only the process itself is of importance, where one can admire the artist's mastery of the most outlandish means of expression. It is an academic art that only incidentally affects the public. It is an art on the scale of global society, which is made possible by *technique* and which is offered for judgment to a public unable to grasp a content or meaning that no longer exists. We must, therefore, examine the over-riding importance of theory, the absence of meaning, and technical ludism. Here we see the oscillation between formalism and meaning with the implied contradictions in this new world of an art as yet ungrounded.

I / Predominance of Theory

From the middle of the nineteenth century the motivations of painters were highly philosophical and idealistic before they became theoretical. One of Hans Rookmaaker's great contributions (*op. cit.*) is to have pointed this out. Thus, Impressionism (and Post Impressionism) seeks to apply the principle of starting with sensory perception in order to acquire knowledge of the universe. The "*Blaue Reiter*" group and later the Cubists searched for an absolute, for the "grounding ideas," for the true reality and structure hidden behind appearances. Piet Mondrian and Wassily Kandinsky sought to transpose reality into constructions that reason could control. Paul Klee expressed the pictorial possibilities and the laws of visual communication in a systematic way. But, this is already passé. Something absolutely new appeared under the influence of theory and *technique*.

Some had the courage to purely and simply proclaim the necessity of absolute newness in art as a function of the absolute newness of our age. Thus Pierre Boulez: "As clever as we have become in trumping the old world … we will not for long elude the essential test: that of becoming absolutely present, of shedding old memories, in order to create a perception without precedent and to forget the weight of the past in order to establish completely new territories." But, can he not see the acceptance of what has produced the absolutely new—the primacy of *technique*?

This art is conceived from an awareness of the possibilities of machines and of the diverse techniques about which one theorizes. Furthermore, he tends to analyze theoretically as a technical process everything that artists have done until now: for example, consider discourse (*langage*) that has replaced what was once the creative act of the artist. But today artists are guided by intellectual systematizations and concepts. It is no longer a question of emotion, of sentiment, of existential message, of a metaphysical experience. ... All of this is largely bypassed and devalued. Art no longer even deals with sense perceptions. The reference point and the cause of a work of art (and not just its means) becomes mathematical. But when I speak of theory, I mean this term in its strict sense as opposed to ideology, utopia, or doctrine. I speak of an academic art that springs from the head and not the heart. I'm not minimizing the highly intellectual role of all art since the sixteenth century, but there is a difference. Art no longer simply seeks the best means of conveying what one means, and it no longer reflects on the rigorous logic of discourse, of music, of design; instead it obeys an integral low of construct with an indefinite perfection of technical means. But, it is no longer anything like popular art but is avant-garde art that is necessarily abstract. The celebrated *Ethérisation de l'image du monde*, by Naum Gabo (reproduced in Lewis Mumford) clearly states the limit of theoretical and abstract art, which simply refuses to reproduce clear and concrete reality. It embraces dematerialization in a design entirely based on mathematics and which resolves the rupture between object and subject, interior and exterior, life and the machine, into a unified image that, although mechanical, nevertheless restores organic realities. The faculty of abstraction realized here goes to the extreme of creating the perfect "symbol of etherization." All of this is conceived theoretically and executed by highly technical means. Sometimes, as often happens at the Paris Biennale, theory dominates the work, but we will return to this matter later on. Now, the decisive importance of theory is found in all the arts. In music, we have serial music, which at first glance is simply theoretical, especially since its extension, beginning with Arnold Schoenberg and Milton Babbitt, among others, to the domain of measure and dynamics. It becomes a type of mathematical composition. But, from another point of view, the systematic search for analogies between visual symbols and sonorous symbols is also theoretical. The same can be said when one undertakes to reintroduce freedom into musical play: here again this is the result of taking a theoretical stance (John Cage). At any rate, we now have a non-figurative music without reference to either history or to the existing body of sounds. The influence of theory is, one could say, greater in music because it is non representational and better reflects the forms of thought. A perceptible order of pure theory emerges from the random improvisations of Andrei Markov or in the use of statistical law in certain works of Iannis Xenakis. It becomes a matter of assembling sonorous objects according to a rule or a group of rules that one has set down. Types of experimental music are also formed on a theoretical basis because the composers create their experiment on the basis of precise ideas. This experimental music is also divided into schools as a function of theoretical differences: concrete music, electronic music, music for tape (Vladimir Ussachevsky). In all cases, it is a matter of creating new sonorous objects, of not taking account of natural sounds or of customary compositions. Music becomes a procedure for organizing new sounds that are totally abstract. Ultimately, one could say that theoretical validity is what makes music. It goes without saying that architecture, like music, lends itself particularly well to this triumph of theory. Since the Bauhaus movement, there has been no end of architectural theory; the construction of buildings has become a product derived from this theory. Architecture does not limit itself to expressing aesthetic intentions or formulating general ideas but also goes beyond the application of technical rules. Between aesthetics and theory it constructs an all encompassing theory of society and of the human being, which is not only a philosophy but is also a scientific blueprint. Le Corbusier has provided a great example. But, currently, Yona Friedman shows to what extent one can push the abstract obsession with theory just as Ricardo Bofill reveals the degree to which theory can be socialized. But, for all of them, a facility will no longer be constructed along pragmatic lines or made according to a cultural tradition. The created work is the exact result of a theoretical concept strictly applied. Mies van der Rohe searches for "the irreducible principles of architecture" as a function of a philosophy of space. We have largely gone beyond structural architecture in a more rigorous and abstract direction. A habitat for man is constructed on the basis of a theoretical preconceived image of man and of his space and his relation to space and also on the basis of theoretical research into forms, proportions, and sensations. Of course, there has always been architectural reflection: the cathedrals were constructed on the basis of an idea about the relation between man and God, etc.. But the great difference lies in the self-conscious character of the intellectual operation in modern architecture. The architects of the Parthenon did not follow a conscious preconception of the effects of proportion on the sensibilities. Now we have the rigorous and total conditioning of the work according to theoretical premises. The unavoidable connection between theory and work did not exist in former times but is now made possible by techniques that overcome all difficulties and obstacles. Painting also has become theoretical. One no longer paints what one sees or feels. Science has taught us to distrust what we think we see. In painting one interposes a theory between the emotion one feels before a color or a spectacle and the completion of the work. It becomes a matter, for example, of multiplying aspects of vision by multiplying points of view. The image should be the synthesis of "all appearances of being." A reality is no longer represented but is replaced either by a will to accumulate at a single point all the possibilities, all the aspects visible or hidden of the object, or by a desire to disperse or break apart the object, because the object, from the standpoint of theory, is merely arbitrary. Jean Leymarie, for example, will say the same thing about Cubism, that it is "the structuring of the real through the metaphor of form, which is to say reality inserts itself in these forms. This is indeed a process of transforming forms." "Cubism is the creation of a new reality because its structuring gaze is perhaps the *a priori* confirmation of current structuralism. Structuralism verifies Cubism, just as the psychology of form, contemporary with Cubism, has verified Paul Cézanne." Along these lines, one sees to what degree art has become a question for initiates and scholars who juggle with the psychology of forms and with structuralism, and also to what extent the discourse on art, meant to explain the theory applied to a given work, has become indispensable. We will take up these two points later on. Finally, and we won't insist on it, we find the same importance of theory in sculpture, for example with Nissim Merkado (very typical among many other lesser known sculptors): the issue involves systematic research on space, the void, volumes, and materials. Sculpture becomes an articulation of geometric forms where junctures are made with steel rods—pure technique but perfectly sterile, the presentation of machines without energy, function, or utility. If we were mean, we could say that we are reminded of Jacques Perret's famous Vistemboir—squares, cylinders, truncated pyramids, complex armatures. I would like for someone to explain the importance of what happens in the void between the volumes that define a zone of tension and action, of junctures, of connections, and of possible events. To take modern sculpture seriously, one must embrace a model of man integrated into *technique*, without which modern sculpture would be nothing more than a play of geometric constructions. But, one can see here to what degree the harnessing of art to *technique* reflects the technical system by taking on its *raison d'être* that can only be explained through a kind of *mystical language* that restores to man an apparent mastery of the object (exhibition at the *Musée Bourdelle*, 1975). Finally, in literature, the new novel, the new poetry are entirely dominated by theory—theory of language, of communication, of discourse, but also theory of the work as work (the novel as novel), *objectalité* as in *Tel Quelism*. ...

    Of course there is not a single theory covering all the arts. In the variety of modern epistemological approaches, in the variety of theoretical tools, there is no single encompassing theory, no synthetization that can bring everything together, but rather there are fragmentary theories, tailor made for each of the arts, and within each art form; theory is further differentiated by accenting such and such an aspect of the dominant theory, which becomes the master key for everything. Whether it's a matter of discovering a world without dream or face, which is inaccessible to a direct reading, or searching for depth beyond direct apprehension, the quest is for hidden origins and for the nature of appearances (Robert Delevoy). On the other hand, for others, it's the search for a rupture in the language (*langage*) of painting and a certain image of man, a search that reveals the appropriate sign, the line, the color, all of which become beings in themselves; their combinations and their relationships constitute a new semantic dispensation. This is a depth that is no longer romantic but is rigorously organized to illuminate the very *being* of phenomena. It is also a theory that exalts chance or randomness (Cage) and accident (Rivette). Spontaneity is not spontaneous: it is the product of a deliberate will, a theoretical reflection. But, the challenge to theoretical principles or to generalities is, in itself, a theory! "True theory should take account of accident: accident is central—provoked, hoped for, and expected. ... Without it the work of art is without interest." (Jacques Rivette). One knows that each figure can be interpreted according to several codes, which produce different meanings. And the only problem is to derive a set of codes that suffice to enable the spectator to interpret the work of art in different ways. But the elaboration of these codes within the frames of chance and accident is extremely rigorous!

    But, theory can be expressed in a total organization of its elements, which creates a feeling of no limits, that Abraham Moles will exploit in his theory of combinatory art. Moles is, without doubt, the only one who has pushed to the limit the predominance of theory made possible by the existence and utilization of the extraordinary technical power of the computer. The point of departure for Moles's theory is simple and rigorous with two factors. First premise: society becomes global and demands an art on the global scale and not on the scale of the individual; it is an art that does not simply reproduce traditional works and that conveys a message from one individual to x number of other individuals. The artist no longer creates alone but rather speaks for a group. Second premise: machines allow for innumerable creations and are docile servants to those in command, handmaidens to the unlimited needs of society at large. From these two premises we derive all the easily deduced consequences that form a rigorous ensemble that is complete and that constitutes a true theory of art based on apparatuses without meaning or sense but just pure form. Aesthetics will furnish the artist with analytical rules, modes of structuring, techniques of programming, and the device, the computer for example. The artist is no longer a simple means of action but rather a producer of the very *concept* of art. In communication aesthetics must dominate semantics. All aesthetic creation leads back to the canonical theme of communication with the transmission of a certain amount of originality where one can enunciate the "universal algorithm of the work of art." The value of theory is given by the extent of its possible application to experimental projects. Reducing everything to its elements, arranging everything according to rules, allows the computer with its unlimited possibilities of combination to become an infinite *omniumgatherum,* on which the art of permutation depends. "The machine explores systematically the field of possibilities defined by an algorithm." Poetry turns into "a game of words." And one can transfer to the word the freedom that has been taken from man. Theory becomes perfectly pure with music, for which there is no meaning to be taken beforehand. Music combines sounds and nothing more. "At his pleasure the musician assembles out of his euphoric fantasies all the sonorous objects drawn from the surrounding universe or from electronic constraints." Pure theory corresponds to a game without substance. Now, the importance of theory should remind us that, the more it pretends to be scientific, the more it prevents and prohibits a consideration of reality—that stupid concrete reality. But, who can live within pure abstraction since we still have a body, since we still find our pleasure in relation to living matter?

All the restrictive principles of selection—the framing of the object, interpretation of forms and colors, choice of tonality, an arbitrary selection of classic poetic structure, etc.—are passé. Today, forms, colors, harmony, words, are molded by the exigencies of their concrete and theoretical uses. No convention prohibits a composer from using any sonority he might need at a given moment and at that moment alone, which will be defined by the theoretical imperative that is art completely programmed (we see this with the computer). Therefore, the determining factor is no longer the object seen, or heard, or felt according to aesthetic convention: it is the device. Thus the musical sculptures of Vassilakis Takis at the 1974 exhibition are based on a magnetic field, or the Ivan Picelj program— a tableau with four thousand elements (1967)—a multiplicity of sound and light engines combined. But this utilization of technical procedures is rarely direct, rarely practical, and is always preceded by a theoretical reflection of which it is, finally, only an application. Thus, one can say that "once the eye has been brought into play, notably by Victor Vasarely, the representation has passed from two to three, indeed to four, dimensions, i.e., it passes from the flat surface of the tableau to forms in space, to movement, and to intermittently diffused sound." (Jacques Michel). In other words, art only exists as such thanks to and through the intermediary of theory. For example, one must make a science of movements, independent of the nature of the objects to which these movements apply, in order to paint correctly or to do cinema that is other than a collection of clichés. The structural disassociation of object and movement allows an assemblage of any kind of movement of any type of object, which leads to pure creation and to the making of art. Another example: art appears through the exploration of a "uniformly dense field of possibilities" that can only be determined by a systematic investigation founded on theory. One could apply the information theory of perception (Moles) to define the rules and codes of combining elements. Information theory, springing out of Gestalt psychology, furnishes the systematic constructions for the play of signs and "hierarchical controls" in both language and interpretation. Theory never addresses content or value of message, never the specificity of signification, but instead offers statistically a possible meaning. The theories governing the activity of modern artists are endless. But one thing is completely uncertain—the value of the theory applied by the artist! Clearly, the artist cannot be held to the standard of pure scientific theory. It is also true in evaluating art of any kind that it is made to be shared and communicated; therefore, one must consider the reactions of the public and of the spectator, of the enthusiast, of the participant, etc. by whatever name you want to call him. And so, one is led to take public reaction as a criterion for the value of theory. One can state directly, like Pierre Francastel, or indirectly, like Moles, that one judges a theory based on the breadth of its applications and on its potential for experimental outcomes. Now it is, by all accounts, unsuitable to measure the validity of things as difficult and abstract as works produced by hermetic theories. We see here a profound weakness and a radical contradiction in the direction taken by modern art, which we will study later.

Let us limit ourselves for the moment to focus on some of the consequences of the dominance of theory. First, this art is tightly linked with all technical possibilities. In reality, theory is not philosophy but, rather, a development of the technical environment. There's no longer an obedience to a kind of subconscious thought but, rather, a formal and rigorous explication tied to the technical mentality that frequently leads to a relationship with the device. Theory is often, but not always, a response to the question: how to utilize certain technical possibilities in art? For example one cannot ignore the capacities of the computer to produce music. According to Moles, there are four methods of utilization: the computer is first of all an analytical instrument for sound itself. Next, computer analysis can be carried out on existing works that will furnish rules and direct a process of simulated composition. Third, the computer can produce totally abstract compositions based on mathematical data. Finally, the computer, through a total synthesis of sound, becomes both composer and performer. Such is the reflective and analytical work of the musician. Then we must recognize a strange reversion: while the artist invents and utilizes the new instrument, the artist is led to the type of music that comes from a computer. We do not go beyond what has always been done, like writing music for the spinet, for stringed instruments, or for the saxophone: the composer writes for an instrument. Fine and good. Now the instrument is a computer that determines what is composed for sound and image, which carries a greater degree of abstraction—one does mathematics instead of musical "composition"—and a greater distancing effect—in painting the artist's touch is missing—that produces a new type of music and painting. Furthermore, *technique* unintentionally contaminates other arts such as poetry, the novel, or cinema: the poet or the novelist begins to compose as if composition was the product of the computer. The poet or novelist is limited to previously established functions or procedure. The artist becomes a programmer. First one thinks, but only a kind of geometric thought—how the spirit of subtlety (*l'esprit de finesse*) is excluded. Painting and music are the products of a program, an algebraic series, a geometric construction. The execution is purely mechanical—industrial or computer driven. As Jean Michel wrote: "the forms of minimalist sculpture are determined in advance, programmed by a creator-engineer who has them produced in a factory: a steel cube two meters on a side painted black can be ordered over the telephone … And, as in a mechanical construction, the minimalist sculpture shows the processes of construction. The same has happened with painting …" The procedure dominates. The result will be inevitably determined by the technical procedure, an expression of the technical era, and problems to be resolved.

Karlheinz Stockhausen shows perfectly the necessity of constant change under the guise of a problem to be resolved in relation to previous compositions and of the possibility of solving the problem through technical means. After music on a single theme (Beethoven, he says, never got beyond two themes—Schoenberg is strictly monothematic), Anton Webern produces two or three intervals in a series of twelve sounds, and now it is a matter of utilizing all possible intervals, "and to develop the series by expansion or contraction toward the micro-tonal level" and "this parametric transmutation of formulae has only been attained thanks to synthesizers." It is a question, therefore, "of transposing all the parameters of one formula into another, treating the rhythm of one melody with the sequence of sound of another melody or with the dynamic curve or sound qualities of a third." "I believe, truly, in new materials in the alpha waves of man, in vibrations that will allow us in a few years to modulate wave and man to enable him to travel beyond our solar world. For, like all scientists …" Thus, Stockhausen declares himself to be a scientist … and only dreams about synthetisms, computers, and vibrational problems to be resolved. But where Stockhausen is radically mistaken is when he believes that he understands—for example in *Sirius*—the "perpetual mutation of nature" by a process made possible by the synthesizer, a metamorphosis of sounds. This has nothing to do with nature but only with what is revealed in the mathematical formulae of physicists.

Art is reduced to a "purely distributional system." The elements of art are no longer definable except by their commutability in the structure of a system, which erases their traditional meanings in favor of their manipulability. For example serial art fragments elements where identities are dissolved and elements are commutable in terms of rules for recombination.

We have a fine example from Jean Ricardou: the writer has "nothing to say" but has to apply a number of procedures in order to write. Similarly, the painter, even when unconnected to the computer, paints *as if* he were plugged into the device. Moles unintentionally relates a revealing experiment. He places a painting by Mondrian next to a painting from a computer program using criteria from Mondrian's painting, which results in two very similar paintings. When questioned, viewers typically preferred the computer's painting, and Moles asks, "Has Mondrian painted all the Mondrians that he could paint?" It's a good question. But this is possible only to the degree that Mondrian's painting, itself, is a type of computer painting, i.e. a painting done either literally by a computer or by one imitating a computer's work. In the same vein Moles compares a computer "poem" to a poem by a real poet. They are indistinguishable, because they are nothing more than a stringing together of words without rhyme or reason with no evocation of music or imagination. "Along walls furnished by orchestras of people. Darting their leaden ears toward the day. Pursuing bodily caresses with lightning. The reaping smile of lowered heads. The odor of sound, etc." Is this a poem? I see it rather as the product of a brainless machine. A significant part of modern music and painting are the indirect sub-products of the computer's impact on the human brain. And so, it is no surprise that one could claim that the computer produces works equivalent to those of the great artists.

These artists have begun by producing what the computer can reproduce. Clearly, the computer can produce works like those of Pierre Boulez, Cage, Joan Miró, Mark Tobey or André Masson (in his later period), because they do exactly what the computer does. By contrast, when one asks a computer to paint like Jean-Auguste-Dominique Ingres or Rembrandt, the result is ridiculous and grossly comical. In other words, the computer can produce the artistic equivalent of much modern art made according to theoretical possibilities developed by the computer. The computer cannot create music, painting, or poetry to equal the works of Mozart or Baudelaire without programming reproductions of these works. Programming is essential to the theory of art based on *technique*. Furthermore, art based on theory, if one cares to understand it, is an art for specialists. It is meant for the specialist who knows the ins and outs of the theory, who can appreciate the enormous amount of preparation, who is up to date on research and intention. Art no longer tells a story nor recounts an anecdote that one can grasp, nor does it convey an aesthetic emotion or an epiphany. Art neither conveys beauty nor pleasure but rather a skill and depth of intention and manipulation without definite meaning. To appreciate these dissonances, these ruptures, these arabesques, to decipher the text in front of you, you require a hermeneutics of modern enculturation bypassing traditional culture. In poetry and in the novel we are given a text where we are not expected to dig out the meaning, which is a question for traditional critics, but rather are given the task of deciphering the text according to extraordinarily complicated methods. In reading *The New Criticism* of Roland Barthes or Ricardou's criticism directed toward the *nouveau roman,* one would suspect a plunge into the world of the Kabbalists. Reading becomes a construction and a reconstruction, which implies a knowledge and an intention comparable to those deployed by the best exegetes of sacred texts. One must devote weeks to a correct "reading" of a *nouveau roman* … We stand in the presence of art of elites for elites. A moderately competent intellectual who does not have the "key" is left out in the cold. I would completely agree with the criticism of Leonid Andreev, when he analyses the *Tel Quel* group in articles from *Literaturnaya Gazeta*, summarized in *Le Monde*, 1967: "Art becomes the privilege of technical experts. One insists on complicating … the work is transformed into a product of the laboratory." One must be in the know to enjoy writing for writing's sake or to enter a totally gratuitous world with no reference to ordinary beliefs, thoughts, feelings, or experiences, a world created by the artist out of nothing, out of dust. (Alain Robbe-Grillet). No question, the artist has always created "his" own world but on the basis of common experiences and on commonplace feelings … and one feels connected to these things. For example Fra Angelico depicts popular beliefs. Now, there is no longer a direct connection between what man experiences and the artificial world the artist offers. Indeed, if there is a connection it is one of *technique*. But, here, modern man is not yet fully adapted to or conscious of what is experienced. Here we have a two-pronged question: is the purpose of art to effect human adaptation to the technical milieu and to produce specialists able to grasp the values of this milieu? Or, alternatively, does the human only experience the technical milieu. Is there not love, a pleasure in nature, friendship, a fear of death, all these clichés, as well? Why, then, do the greatest examples of modern art reject this common experience? Why, then, must traditional concerns of art re-emerge in the form of song … The art of the specialist, of the technician presupposes an explication.[1] Modern art is nothing without the explicator at hand who will explain such and such a trait of Mathieu or such and such a conjunction of colors in Masson … Without the expert we are lost. The viewer needs no explanations for seeing the works of Vermeer or Renoir, but Hartung or Dubuffet can only be appreciated through the commentaries of Dorival or Coreau. Bernard Dorival situates Hans Hartung's abstraction between the impersonal formalism of constructivists and abstract expressionism. "How do we understand the rejection of representation by great abstract artists? Instead of representation, abstract art substitutes forces for the traditional conventions of painting whereby Hartung aims for the truth of an interior world, an extension of his own ego. For Hartung abstraction is the basis of true painting." (Dorival). Explanations carry an ensemble of implications. First of all, we should note the close pairing of subjectivity with theoretical art. This conjunction is a result of formulating a theory and the occasional application of mathematical principles that the artist claims (or experiences) to be the revelation of his most inner self beyond the space of form that leads to a purely interior world appearing in the negation of the exterior world, its denial and its annihilation.[2] This may appear as a spiritualization of art; when a musician programs a computer the musician is viewed as being in command. Mind has finally conquered matter. Mind has liberated itself from cultural models that restricted spirit with the contingencies of motors and means. Once again, Adorno has viewed things correctly: "Works of art, which have rightly devalued the fascination of what we see, nevertheless need sensory input if they, in the words of Cézanne, want to come into being. Ironically, sensorial devaluation occurs despite an incessant and uncompromising demand for spiritualization; the more works of art move away from reality to be sublimated, the more the gulf yawns between hovering spirit and its sensorial base. The primacy of coherent ordering suffers a reversal: the domination of the spiritual over the material produces a loss of spirit, i.e. a loss of immanent meaning." Thus, according to Adorno's unassailable logic, the modern art movement ends in a denial of meaning to a work of art and an elimination of the possibility of creation: ultimately, pure subjectivity can no longer and must no longer create. Silence and absence.

But, we make two other observations on the relation between subjectivism and theory: first, subjectivism is increasingly "pure," more extreme and uncompromising. Hence, modern art denies the possibility of communication; even the expert who has grasped the theory is caught in the essential paradox: nothing has been communicated with such sophisticated means other than being-in-itself, which is as incomprehensible as God himself. Modern art must be approached with parables and metaphors, which, ironically, are no longer symbolic. And, we have seen why: the technical milieu sterilizes the possibility of symbolization. In fact, modern art denies parable or metaphor and claims merely to be, being-in-itself. That is all.

There is no subject in modern art but merely an artistic intention that is so recondite and sophisticated that it is impossible for the listener, the reader, or the viewer to understand it. Here is one of a thousand amusing examples:

A glass of water on an open umbrella entitled *Hegel's Holiday* by René Magritte. In a letter the painter himself replied: "My last painting began with the question: how to depict a glass of water in painting in a way that is not commonplace, fantastic, or weak, but let us say, ingenious, without false pride? I began by drawing many glasses of water … always with a distinctive line on the glass … this line, at the end of 150 drawings, splayed out … and then took on the form of an umbrella … This umbrella was then placed inside the glass … and, finally, underneath the glass. This was the exact answer to the initial question: how to paint a glass of water with genius? I then thought that Hegel, another genius, would have appreciated this object, which has two opposing functions: to repel water and to contain it. He would have been charmed or amused, as on vacation, and so I call the painting *Hegel's Holiday*."

Clearly, the explanation is indispensable. But, one can also add that, in a painting so clearly figurative and with a "subject," the effect of abstraction and symbolization is as great. Would one need so much explanation? I would say, no, because this is in no way a purely individual work according to the process Magritte describes, but it is a work of art still connected to a common symbolic system, which no longer exists. The artist is left with a purely gratuitous, elusive, and individualistic game to which modern *technique* condemns aesthetic creation.

But, as one last remark, this pretense of subjectivity unveiling an inner world really veils a total ignorance or an unconscious denial of the control of technical means. It is astounding that these extremely theoretical artists do not see that the theories they formulate reaffirm their dependence on *technique* and that the pretense of free expression, of pure subjectivity is merely a conditioned response before the consequences that this would entail. Whatever the case, discourse is increasingly present beside the work claiming no longer to be discourse. Explication becomes central when the work no longer stands alone or is meaningfully explicit. Thus, there is an overabundance of disquisitions by the author about what he is trying to do, about his intentions, his feelings, and his experiences, and also by commentators who spend their lives deciphering the obscure signs left behind by these latter-day Neanderthals who are the artists of Saint-Germain-des-Prés. Certain people may appreciate the flourishing of doctrines and experiments, seeing them as "a turning point in the relationship between art and society," a way of accelerating the transformation of public taste.[3] What a surprise when one finds the disquisition has increasingly replaced the work of art. I can take a blank piece of paper and in a trice trace out an arabesque, and then go on to explain on page after page or hour after hour on television what I have done, how I have done it, and what it reveals. And so forth. Conversely, when a poet or a novelist, who will also hold forth on a puzzling text or on asinine output, the text is substituted with explanatory iconography and the substitution is complete. Philippe Sollers replaces what he does not write by an indefinite commentary, a disquisition on what the work that is not should be. Furthermore, Roland Barthes offers an excellent justification for this: what is produced, according to him in *Critique et vérité*, is a permutation of functions. The critic and the writer combine "to confront writing," and what is left is neither critic, nor novelist, nor poet but people coming to grips with writing. Writing in and for itself. Without realizing that this mad rush is directly caused by the technical system, which bedazzles and hypnotizes us with the means to an end. The means alone exist and matter, and that is writing. This is nothing but *technique* decked out in all the current magnificence of *technique*. But, of course, one is unaware of this. What decadence. Again the weakness of these analyses, so subtle, so elegant, so dazzling, etc. appears. When Philippe Sollers sees in his approach to writing a new direction that is related to surrealism in its attempt "to displace the frontiers of reality …" he simply does not see that the displacement has already occurred! *Technique* has displaced the real, which has substituted a new reality for the old in an approach to writing that is reality's pale reflection. But theoretical discourse on art is as disappointing as discourse on art with a message. The latter flounders in the infinitely banal. The new approach is marked by an astonishing and confusing pathos. We hear serious words about "writing microcosmic music," about "a rhythmic and vibratory experience of the world, manifest in tension with color itself." (Delaunay). The invisible is called to the surface of the visible by examining oneself, and a phenomenology of the invisible should be based on "Gorky's scriptural induction." "The fall into the abyss" argues against "the fissure where my desire holds sway." We have a completely private language of a fissure, a juncture, a slippage, a gap, a decentering, etc. This discourse reveals, unintentionally, an elaborate and complete theory that empties the work of meaning. Spiritualization carried to its logical extreme means the sterilization of the spirit. The hypertrophy of technical means signals the disappearance of a purpose behind those means. The conjunction between man and the technical means at his disposal indicates that communication is broken by a burgeoning of those means. We no longer witness a process of aesthetic communication characteristic of an artist working in solitude who becomes the true medium of communication by renouncing the standard forms. Artists once spoke to men thanks to their solitude, which no longer exists. Artists are now engineers who manipulate devices in theoretical applications that reveal a total absence of signification in the discourse about this art. This brings us to the loss of meaning, the loss of sense.

II / The Loss of Sense

Art no longer has either referent or reference. Until now, art corresponded to beliefs based on pre-existing values that guided all that was said and done. These values have gone by the board in a free floating art with no reference, art unleashed, if one believes what artists claim. We should not forget, however, that values have disappeared and that *technique* reigns supreme.

With the fine insouciance of genius, John Cage declares, "One must be disinterested, accept that a sound is a sound, abandon our beliefs in order, forgo expressions of sentiment and our verbose aesthetic baggage. The supreme goal is to have no goal. That puts you in touch with *nature*." I could add similar declarations from writings or interviews of hundreds of painters or musicians. What can we take from all this? First, the recognition of brute facts and man's abandonment of imposing order upon them. Thus, a sound is a sound. The musician produces sounds without order or feeling like the squeaking of a closing door or the backfiring of an engine. The musician becomes a thing. He abandons all that he has done in the past; he no longer embodies a place in an environment; he no longer attempts to impose an order—illusory or not—that is essential to man. Thus, we are in touch with nature, a curious personification and an abuse of language, because "nature" itself is a construction of man. But, for these wise men, these notions are not incoherent clichés. The goal is to abandon all that man has accomplished and to completely embrace nature and finally to renounce all traces of artificiality. There is nothing new here in the way of hope, doctrine, or will. One cannot truly see in this situation what a musician intends to do. Why does John Cage give concerts that are not in nature, and why does he claim a need to be revolutionary when there is nothing natural about revolution? Enough said. To embrace nature one must do so without a goal. "I am a force that moves," as Hugo would say a being without intent, aim, or objective. Nature is a combination of forces, accidents, necessities, and blind drives. John Cage settles, unintentionally, the question of finality or intentionality embodied in nature, but this problem is far from resolved in spite of J. Monod. It is a question of choice. Let's grant Cage's choice. How is it that he does not stop theorizing? Why did he adhere to serial music and then move on to the theory of chance? Why does he not realize that in so doing, he is no force of nature but rather an expression of the force of *technique*? It is *technique* itself that is without finality and goal, as I have shown.[4] It is *technique* that is a "being-there" and nothing more with no need for meaning or value; it is because it is. It imposes itself in two senses: as an overwhelming and undeniable power of efficiency and as an absolute that results from man's beatific embrace of efficiency that is *technique*. John Cage believes that by eliminating goal and sense he is in touch with nature, but in truth he is in touch with the technical system. From that point on there is no conflict between his proclamations and his theories. In fact, in his person, in his discourse, and in his work we see the unification of extreme theorizing with the absence of sense. Art that is theorized is art without sense. But, this is a matter of degree. One notes, indeed, the loss or the death of sense in the works presented to us. There is no point in looking for meaning in them. This happens because one is ready to take the text, the painted canvas, the musical sound as an object, as nothing more, which is a thoroughly technical attitude. *Technique* is not concerned with the content of a message but only concerned with the structures of its presentation and of the object existing as such. One can ask whether this attitude is a result of the fact that we have a new concept of art (this is clearly the claim of those who make Theory and who are unaware that the attitude derives from *technique* itself). Or, alternatively, we forgo a search for meaning because there isn't any, and there isn't any because the modern artist, painter, novelist, musician, has basically nothing to say and is perfectly hollow and empty. He produces text and color, but he puts nothing into them because they are nothing and are perfectly sterile, which comes from the technical milieu in which they find themselves. Theory then becomes a substitute for the absence of being. One combines algebraic symbols because there is nothing to see, and one sees nothing because one is plunged into a milieu that does not allow profound or mature experiences that suggest consequences for what we produce and understand. An aesthetic emotion does not arise through an active involvement with the work. There is no meditation on the meaning of the works (hence, the success of the bluffs of theoretical gurus and their transcendental meditations). There is no longer a distancing from the work because we are seized by the imperative of immediacy. There is no longer an attempt to share meaning and value because the imaginary museum, produced by *technique*, has succeeded in persuading us that there is nothing left to say. And, therefore, let us remain silent. Let us write or let us paint, clinging to the bare act of writing or painting. This attitude perfectly embodies *technique*. We now see it taking full effect. Everything has been said and the train has left the station. What a strange prophecy when in that time so little was known about what had already been said. Currently, it is exactly the same. And every artist is barred in advance by this co-option of messages and meaning. What could he add except color, raw matter, and sound? These at least exist.

At the same time, the disappearance of the artist occurs with the rise of process. A case in point would be William S. Burroughs and Brion Gysin in *The Third Mind* (1977), expanded in *Oeuvre croisée*. On the one hand the two authors are transformed into a third who eliminates them, and on the other hand, they exploit techniques meant to deprive them of all power in constructing the text: for example, *cut ups*, a systematic and mechanical form of collage (all the words of a sentence are reversed according to a mathematical law applied independently of the authors). The aim is to "deny the personality of the author and his control over words," to remove the writer in his capacity as a subject with something to say, and to explore the possibilities of language in order to escape as far as possible the writer's alienation. If we leave aside the internal contradiction of this procedure—why try to escape alienation if man no longer exists?—the pure technical process of such undertakings comes to the fore. The autonomy of the subject is lost thanks to the use of a mechanical process (which is clearly a characteristic of the technical world), and what comes first, what counts and matters, is to work out all the possibilities of a given instrument, in this case language, and this is precisely what happens in all technical processes. And if one claims that language is not an instrument (bringing to bear all that one knows so well about linguistics and structuralism), I will say that it is precisely a thoroughly technical form of discourse that becomes a metaphysical instrument while at the same time denying all metaphysics. All of this is characteristic of the technical mentality.

But how may one remain in check with this understanding? One must justify oneself. Hence, one declares (and this is a new step in the theory) that *there is truly nothing "to be said."* Or, that above all, one must say nothing and mean nothing, because, metaphysically, there is no meaning or sense. There is no longer any possible *intention*. And even though this art may be perfectly subjective, there is, at the core, an aspect of subjectivity that is condemned, namely intention and the search for meaning, reason, and thought. The artist must strive to say nothing in a totally negative aestheticism, which implies a final stage. It is not enough that there is no meaning, because meaning can always leak out like the leaks in poorly sealed butane canisters. But let's be clear, for there must be no meaning. Meaning is dead but maybe not entirely, in which case we must kill it. Thus one passes from awareness and impotence to a decision to eliminate, which itself derives from a metaphysics masquerading as pure theory, although it is not so pure. It is a metaphysics that perfectly follows from the technical system that grows from the obstruction of man's moral beliefs and values. We now arrive at the last stage of a complete isolation by what should have been a means of communication, but a communication of *something*. Technical art is offered as a wonderful network consisting of telephone cables of every color, of every capacity, of coaxial cables, and of automatic switchboards, but wait, there is no one at the mouthpiece, no one to listen or to respond. Simply, the public is called upon to admire the art of the engineer and the skill of the workers who are able to produce such bundles of wires. Clearly eliminating meaning is not achieved to the same degree and is not expressed in the work of all artists with the same force—mediocre and lame in some and desperate in others. Some produce a little of whatever out of laziness and rely on the theory of a loss of meaning as their excuse. Since nothing has meaning, why try? This conviction provides an absolute security and an unquestioned certitude with no content that is quasi-religious, and this is why when one speaks of meaning in such a milieu one is ostracized. But, this attitude is not just a result of laziness. There is more. How can one be unaware that art that produces meaning attaches itself to a sequence, that meaning is not an unexpected outburst, but rather a participation in an ageless quest, hence it is the handmaiden to the "affirmative essence of culture." (Marcuse). In other words, *all* meaning is by nature conservative and reactionary. In order to shatter cultural conditioning one must challenge all meaning, whatever it might be. That alone is eminently revolutionary. Furthermore, so long as art refers to the representation of an object, it confirms the object in its significance, but that again is conservative! Hence, refusing aesthetic signification accompanies a denial of all internal or external similarities or relations. One more thing: we are familiar with a tendency of the arts to fuse and to combine with each other. This also relates to the death of meaning: as long as there are specialists an ideal of order and harmony obtains. Visual and auditory criteria are held separate. Correspondences are subtle but harmonious and not muddled. Indeed, harmony and order are tokens of integrity and the possibility of meaning. But, if one denies meaning, then it is absolutely necessary to introduce disorder into the arts. And art, which can no longer be symbolic, will find an expression of its new nonsense in the "montage." Gluing bits of newspaper onto a painting (and all the developments that this "technique" produces) is not a new form of art: it is the negation of meaning in art by the injection of fragments of everyday reality. Exterior objects are no longer imitated, designated, or symbolically attached to or immersed in a meaning. No: a piece of newspaper is thrown in the face of the observer, framed by fragments of color. Art then is involved in something completely foreign to it while it no longer fulfills the signifying function it has traditionally held. "Art becomes essentially an object among objects; it becomes something whose nature is unknown." (Adorno). It is, simultaneously, and by the same token, the art of no one. "The I is an other. Poetry will be made by all." Sadly, the community of all has become the computer. It is clear that we always come back to the same point: if art is essentially the bearer of meaning, then the computer has nothing to do with meaning. If the computer imposes itself on us in all its splendor, then art ceases to have meaning, because the computer can do everything (although that remains to be seen, but let us grant it for the moment) except to create meaning for the stranger who is man.

But this fundamental dilemma taken to its logical extreme has already been posed by mathematical music. The transformation of expression into matter, mathematically radicalized, has already challenged any form of expression. Rigorous, logical construction petrifies art. We move toward methods for organizing sounds, colors, and forms that are totally abstract and which, of course, exclude all intention or all emotion and are only valid to the extent that their aspects have been well chosen (naturally no perceptible sensory relations are possible). The instrument, the apparatus, totally dominates the process that traditionally it served. A final goal is no longer at issue. To this point we have mainly talked about painting, music, or the novel, but the process of dehumanization is even greater in the theater, for here the theater is reduced to an absence of text, gesture, and mime. That's all. Even with Rorschach, of whom we have spoken, events happened on stage. With the Teatro di Marigliano and Della Maschera (presented at Bordeaux in October, 1973), we have total nonsense by way of mere images, which produce a new "theatrical poetic." "A piece of moving material fragmented and completely incomprehensible." A set of sequences, discontinuous fragments revealed in successive illuminations, disconnected phrases, and minimal music, fascinating in a way, and consisting of a few indefinitely repeated notes, which evoke a possible continuity through the reappearance of fleeting motifs …

The idea that the word is the essence of theater is rejected (for the benefit of the gesture, the grimace, etc.). The idea of language as a tool for communication is likewise rejected with distrust and doubt; language is derided. "We renounce the superstition of a theatrical text and the dictatorship of the author," says Antonin Artaud. Language is inadequate and hence subjected to ridicule in order to emphasize non-communication, the bankruptcy of words, but how do such playwrights not realize that they are merely mimicking *technique*'s attack on language? It is the technician who says above all others that spoken language is inadequate, imprecise, and content free, and is purified by a drawing or a graph (in theater gesture, pantomime, and bodily movement do the same thing!). The technician challenges "empty talk" for the benefit of algebra. And our artists, enslaved without knowing it by these claims, commit artistic suicide and murder Art, communication, and the word in a bow to the great organizing forces of our society. The brave young people who think they are nonconformists by doing the theater of bodily expression are only ancillaries to the triumph of *technique*. Only the word, bearer of meaning and communication, is truly revolutionary and able to call into question the technical Empire.

If one looks for meaning in a poem or in a painting, one will find nothing in the painting or text, which must be seen as that which is, because painting is simply a matter of painting (and not of painting something) or because poetry is simply lining up words and not formulating ideas. Edouard Manet, for example, according to Daix, "exalts in the act of painting without needing God, nature, or other correspondences; his imagination is *in the painter's touch.* Hence, for him to identify with literary, historical, or social significance in what he paints, would be meaningless." Daix hits the nail on the head with the following: "Manet expects from painting a simulacrum of painting just as Baudelaire in his poetry would seek a fabrication of morality." Today, Manet is the winner. Everything has gone his way. The image, the sign, have become valid by and for themselves. The painted object, reinvented by painting, has its existential worth only it itself. There is no need to refer to anything—neither to the artist nor to an idea. Everyone currently knows that signs (sounds, words, images) carry elements of information independent of the intended meaning conveyed by their medium: the task of *technique* is to continually refine these signs. Modern Art works toward a radical break between independent bits of information characteristic of signs and the message, which ultimately must be excluded. Kandinsky said that, "… a dot can sometimes express more than a human form," which follows from someone who sees the human form as nothing but an ensemble of lines and dots. But, according to this logic, Kandinsky is off base when he says "expresses." To whom and about what is this expression? To nothing and to no one!

Modern art must tend toward the goal of emitting "signs without meaning, proposing a new meaning for artistic essence, one that is totally abstract and simply a code of rules." (Moles). But we can take this a step further. Ultimately, "the search for meaning excludes an understanding of the work by the spectator," and then: "The concept that the artist creates forms but not meaning," becomes necessarily an anathema to meaning, "which must be abandoned willingly and forcefully." The signified itself, in well-known linguistic discourse, loses its consistency. The signified is not a "reality," and, because Saussure writes that, "the linguistic sign unites, not a thing and a name, but a concept and a sound-image," one can happily abandon the concept in favor of the sound image. If one desires to paint or write with the least bit of intentionality; well, then kill the desire. Literature and painting are "productive activities free of all antecedent meaning." (Ricardou). Therefore, we are left with a simple technical praxis that only fully achieves its goal to the degree that it is void of meaning. The "dogma of expression" (expressing an idea, or something, etc.), according to Ricardou, is evidence of total incompetence. A modern artist must avoid this. Ricardou[5] is probably, in effect, the prince in this struggle against meaning with his elaboration of an entire system—a text in its pure state, without author, without reader, without a literary character. Word and syntax are the sole producers: they function, they produce, and the novel and the poem are "machines" (clear technical dominators). What is the machine but language and text? However, they can only function this way if the machine has nothing to say. The writer has nothing in mind. We are familiar with Ricardou's violent attack on Sartre, who made what was to be said more important than the means of saying it, petit bourgeois ideas according to Ricardou. "The Sartrian doctrine is only a slap-dash generalization from one particular case; if at dinner someone asks for some bread, the object itself is privileged over the signifier." And here we have a use of language for specific ends. The essential and the true are the text itself, a text that is as yet unwritten, but which *ec-sists* as text in an embryonic state and which demands to be born. The imperative to write the not yet written fuels the writer with the desire to write. However, the text has no content but is merely an imperative of pre-established linguistic functions (in other words, *technique*), which gives birth to the text. In this universe man in all his dimensions is annihilated. There are no longer characters in the novel. "Traditional characters are replaced by grammatical constructs resistant to all appropriation." And the author, of course, is a mythical character, constructed from the specifics of the text. The author is the transmitter of orders and the executor of a calculus. But the coherence of the works is suppressed along with the author. There is no reason to regard pagination because one text blends with another: no text can really belong to a single work, since in all cases, there is no coherence of meaning. The *raison d'être* of the text is the text itself. One puts some initial propositions into the mechanism and then all develops automatically through a series of cycles and sequences, but who is the "one?" The reader? Definitely not the reader who merely applies the techniques of reading. The author is eliminated for the sake of the text and the self-generating role of the word. The reader is eliminated for the sake of a set of techniques for reading, a solipsistic reading. The reader must be armed with infinitely complex techniques to decipher the absence of meaning in a text and to see its *ec-sistence*. The levels of reality in a text are only revealed in the application of these techniques. To that end, "the signifier is given a decisive and productive role." In order to give free play to these textual machines and techniques of reading, one must pursue, root out, and kill meaning. There are no better examples of the subordination of thought to the technical process.

At this point, in spite of the system, we have an unavoidable difficulty: modern art denies conventional meaning as it exists in our society while, at the same time, embodying that meaning as *technique*. It challenges meaning by denying all meaning other than the means of *technique*, and thus it is impossible to totally exclude meaning despite all its efforts. The loss of meaning proclaimed by the artist and theoretician, as if they wanted to destroy what had constituted art, is never the last word. The Theater of the Absurd (Beckett, Adamov) does not identify what is usually understood as meaning, but, by challenging its own meaning and consistency, it reconstitutes a new internal coherence and, unwittingly, a new meaning. When we stop evaluating *technique* with a natural or human model, *technique* becomes rational and meaningful, a meaning which is neither more nor less than *technique* itself. The conflict between natural meaning and technical meaning is perceived as a kind of suffering resulting from the domination of *technique*. This conflict leads to the problematic of computer art that tries but fails to mitigate that suffering. As traditional meaning and sensibility is challenged by computer art meaning, a new dogma and morality simply takes its place.

There is still the possibility of taking the signs apart. All is linked to their reality and existence, but negation is around the corner. When an orchestral concert consists of destroying with hammer and hatchet the instruments of music, when one holds a seminar on *Destruction as an Artistic Procedure* (University of Oregon, 1966, cited by Mumford) and, when, in a class on visual semantics taught by Morris Yarowsky, a student throws a cake on the floor and another smashes a television screen with a hammer, we are in the moment where the sign itself is illuminated. This leads us to the realm of non-art, non-work, and to the place of pure absence. But this is not an effect of *technique*, but is simply the prolongation, the result, the consequence of the denial of meaning, which springs from *technique* itself. Now, this denial of meaning is the denial of art itself. Here we are in the presence of a basic non-demonstrable alternative, which seems, nevertheless, to derive from the relations between art and the society that produces it. Francastel is correct when he writes that art is not a pure game of imaginary speculation. It is impossible "to add or subtract at will certain elements from an organically constituted unit. One cannot define beauty as a power outside the object; the object is always a product of the total man. It is completely arbitrary *to deny the aesthetic object this participation in total meaning.* Art is first and foremost a unit of information. In a work of art one always discovers modalities of action that are absolutely different from those of *technique*, which is only a social institution." If there is no subject, no intent, or no transmission of meaningful information beyond mere operation, there is no art *or at least what has been usually meant by this word.* If this is the case, the movement we have just analyzed is the denial of all that has been considered art within human history. It should, therefore, be no surprise that art becomes the destruction of the sign itself after a history of idolatrous reverence, which manifests the irreconcilable difference between art and the technical system.

III / Technical Ludism

For those who believe that art should carry a message, art was considered a game with revolutionary import, but, also, for the theoreticians of formal art, art was simply a game. In reality, for both groups, the instruments of *technique* become the tools for this play at being technicians. The idea that art is a game coincides with theory and with the absence of sense/meaning. It is a game like chess or bridge: the act of play has no meaning beyond the play or the distraction it provides. One plays because one enjoys a game that fills time and prevents thinking of something beyond the game and because play brings a group together. Art is the same. But, at the same time, a game requires the enforcement of very strict rules and principles and a strictly determined praxis. The discipline of the player follows a set of rules and complex imperatives to create something both personal and noteworthy. We see the conjunction of the absence of meaning with theory, which neither explains nor justifies itself, because art is a game.

But with these theoreticians of art as a game, there is a slight modification; they say in all seriousness that art is a game in itself and that it has always been so and nothing else, that man is a player, and that art is simply an expression of this trait. Therefore, to be taken seriously, one must avoid metaphysical generalizations about Art as if one were able to form a universal definition of art and as if such a complex activity and phenomenon could be meaningfully designated across history and culture. The greatest error is to conclude that in art today, "We understand the nature of the phenomenon infinitely better than those of the past; until now people were misled in not seeing art as a 'game' for all people in all times." However, this is to conclude that art is and always has been what it has become in our own time. Art has been reduced to being only a game by the efficient utilitarian ordering of technical society. But one must also realize that, by affirming this audacious and revolutionary claim, our theoreticians are exactly adopting the bourgeois attitude toward art, which, as we have already noted, formed a part of the pleasures and amusements of that class. By denouncing the spiritualization and idealization of art by the bourgeoisie, they are merely condemning, unknowingly, their own reverence for theory. Indeed, the only value recognized for art by the bourgeoisie was that of play and entertainment. Moles is quite right to write of his theory of art as a "game of permutation and combination." The consumer is offered a host of combinations among variable elements,[6] and the excitement necessary for artistic pleasure is created by the immense possibilities for exploration. This art is a game of possibilities, a mode of aesthetic exploration, accomplished by putting meaning in parentheses and by not expressing or expecting any value beyond the play itself. Play, like life, has no meaning. But the consumer must be steered toward the most interesting productions by means of complex rules and "filters." The game is only a game, however, if one respects a set of rules and constraints. One can no longer expect, in a society so wonderfully technical, to produce a work comparable to those of the past! "The poet, as the etymological titleholder of linguistic creativity, no longer writes traditional poems (except by mistake) but, rather, expresses better than ever a society that discovers art as play. He sells himself to the builders of syntactic structures and discovers with the 5+7 method, dear to Jean Lescure, that meaning is the most recalcitrant of Gestalts to all the distortions that the demon of the game can produce, and so, like Dufrene, he dissolves poetry into phonetics …" (Moles). Rules and constraints are produced by mathematical and technical imperatives. But, of course, in order to attenuate the insipid and foolish nature of the rules, one is informed that man now plays "like the gods of Mt. Olympus." Of course, thanks to machines one can pretend to be Jupiter. The products (one could not call them works of art) are little more than other popular games. They are equivalents of crossword puzzles, rebuses, puzzles, word games, riddles, amusing problems, etc. The explanations of Ricardou on the texts by Roussel, for example, show that the latter was indeed proposing riddles, cryptograms, and guessing games. To dismantle a text one teases out the correspondences between " key words." When the texts state that the boat breaks up, one means allegorically that the text breaks apart; and when one writes of black men and white men, an allegory of paper and ink is formed that designates writing. The reader is invited into the game that Edgar Allan Poe makes in *The Gold Bug* to read for the hidden text "deciphering a writing within writing." The author writes a text that is to be understood as a rebus or a cryptogram; the rules of the game are not given, the first part of which is to find out the rules and the second is to apply them. Or, again. the painter presents fragments of a puzzle, and the viewer's job is to establish an order unlike that of ordinary puzzles where meaning is represented. Robbe-Grillet in his *Project for a Revolution in New York* (1970) moved the novel toward a game of cards shuffled at will. He offers a series of images tangled in time and space, stories exploded that must be glued back together. And Julio Cortázar announces in his title *Hopscotch* (1971) the intent of the work. One begins his book anywhere. The "textual blocks" follow each other without any relation. Nothing happens and nothing is explained. The reader is called to modify the text as he sees fit. Each moment of the narrative is independent and cannot be linked to any other given moment. Each literary character has a double or a triple. Any action and any literary character can be created by any other action of any other character. There is no causality. Simply, we have lottery balls, nice and round, that are arranged as seen fit in such and such a compartment. Each reader can make the novel he wants. "Volez-vous jouer avec Moâ?"[7]

Your mood determines the meaning of the work. If you are down-hearted, so is the work. If you are high-spirited, it's a great flight of fancy. "Finally, one has found the means to make the reader a participant. The spectator or listener is moved to action and to construct for himself the work with the materials the artist has wisely chosen, polished, and laid out in the proper manner." As in architecture, according to Yona Friedman, the dwellers must form for themselves their own environment according to their taste but, unfortunately, that can be accomplished only in the framework of a highly technicized world: the dwellers can remake everything, challenge everything according to their taste, except the marvelous possibilities of the technical system. Once again, we are left with a décor. Serious matters lie out of reach and suggest the function of art dictated by the bourgeoisie. It's a Roman-puzzle that must be treated as such in order to find any interest; beyond a puzzle there is no sense or meaning. You will find no beauty, depth, value, or truth beyond this uplifting puzzle that energizes and motivates by the complexity of language. The painter offers images of Op art in which, "the artistic element is the subtle play of perceptions oscillating between layers of meaning produced by the repetitions of circles large and small, squares and general patterns." (Moles). This belongs to the games of "optical illusions" that appear in all the magazines, and in aptitude and ability tests. We also see games based on word and syllable associations and games of punning that are quite popular. One reaches a very interesting juncture: with reference to typewriters and computers (the computer itself that *composes* a text). Moles affirms that, if discourse is to be given meaning, it will produce an unacceptable *bête noire.* But if one takes meaning away from discourse, if one sees discourse as merely a series of sequences, such as "word associations functioning as tools," as "pieces in a montage," one obtains a valid text. Then word play flourishes and one will begin to write things like, "A knotty problem—not a problem." Or, "Would it be etiquette if he ate a kit?" Or, "A pair of peers." [What follows are a series of sophomoric puns that cannot literally be translated. Translators.] You can go on with these forever.[8] I am struck by the poverty and idiocy of these puns, which are taken from the *Almanach Vermot*. As for the play of art, we have ascended to the level of current television game shows and the spoonerisms of early twentieth century vaudeville. This art play is on the level of the comic trooper of times past. After having constructed a very complicated theory with the utmost scholarship, and after having written the deepest reflections of writing on writing or on language, one is finally left with nothing more than ridiculous mouse talk. It is clear why these same critics ardently challenge the very idea of a work of art. Clearly, nothing of all this grandstanding has duration or permanence just as crossword puzzles done on a train are not worth saving any more than episodes of a television game show hosted by Guy Lux. Now, this is the level of our descent. These are crossword puzzles for highly qualified intellectuals or Guy Lux programming for dyed-in-the-wool aesthetes. But since we are considering a game, there must be interest and amusement. If one plays, it is to escape the grey, dull boredom of the everyday. But what is proposed is not at all playful. One is invited to agonize over incomprehensible material. This is not play for me. I will even claim that I am more bored and frustrated when I read Robert Pinget or Claude Simon, and when I hear Boulez or Victor Barbeau, and when I look at Vera Molnar or Mondrian, than when I am in the métro.

This is not an issue of intellectual difficulty. I can read rather difficult texts, more difficult than Roussel; it's really a question of boring exhaustion and emptiness after an initial moment of curiosity. There is definitely no playfulness here. If the theoreticians want a festival, let's go to the circus instead and chuck this "theoretical-art-play." It has achieved the limit of its failure. Who could experience pleasure in this game? Only a specialist in the game? Ricardou will gaily play with pieces provided by Simon, but with him alone. The spectator who is not a professional in modern literary affairs is bored and passes by with a yawn. No, we do not want to play your little game. When one sees the results of theoretical art-play—and there are always results—one is bored to tears by an explanation of how they are obtained, and one is faced with the famous questions: do they paint that way because they are incapable of painting? Do they poetize in that way because they are impotent? Sadly, yes. And the gravitas, the depth of the theory is there only to hide these incapacities and impotence. Technical society renders the creator of art impotent. He can only be an entertainer. But technical society infuses the artist with its own characteristics. An extraordinary complexity, a prodigious grandeur of means, an intellectual and technical power without equal, gather to produce a labor of the negative.

The most perfect theory of organization produces incredible disorder. The most advanced computer never succeeds in producing for the artist what it theoretically could produce. The production of gadgets is multiplied to infinity, and it's not due to chance or to the ill will of nasty capitalists, who create the useless out of the useful, allegedly for the good of the people, but it is due to the nature of the system itself. So long as the system remains intact (which is implied by the growth of *technique* itself and not by some erroneous political will) the technical system will continue to produce gadgets. Theoretical art is one of the gadgets very useful for the survival of the system and for the incorporation of the individual into that system.

IV / The Oscillation Between Formalism and Meaning

We now enter another domain of contradiction in modern art. We have seen art that claims to carry a message, but we have also seen art as pure formalism. We have already addressed these contradictions in describing the ruptures in art, but now it is possible to indicate certain new aspects of this situation.

We begin by granting formalism in technical art. One begins by eliminating meaning/sense and then by seeking what can only be pure form. One searches by combination, which is the nature of the technical process: one seeks all the possible combinations among certain elements according to certain rules. That's all. The search for pure form and linguistic structure and for sets of sounds and words is justified in the form of research, which is characteristic, again, of the process of *technique* in which all content is eliminated. Therefore, all can be reproduced, as we have seen, without human intervention. When one arranges sounds—or combinations of sounds, for example like those produced by shortwave radio—how can one not take this as formalism? It is very curious that in the fundamental debate of our time about what is History, art as a whole has swung to the a-historic, anachronistic side, claiming to summarize and to concentrate at one point and in one instance all the artistic moments in time. But, between History, which carries meaning, and *technique*, without meaning or sense, there is an insurmountable conflict. The result for art when the choice is pressed comes down to formalism. Of course, I understand that all modern artists challenge this term and reject the very idea of formalism, but how could it be otherwise? We are not writing about a *school* but about a *reality* in this production of art. Furthermore, with the primacy of form the work of artisans and the work of mass production are reconciled. Individual form invented in a way that is not precisely individual is not opposed to mass-produced form. When one seeks a common thread among the so many diverse artists in painting and sculpture working for the past twenty-five years, we find in the steel geometric forms of David Smith, the tortured polyesters of Robert Mallary, the Paleolithic stones of Ralph Stackpole—and I could enumerate works forever that seem completely divergent—their commonality resides in the search for pure form.

Art that claims to be informal is really a quest for forms, but the forms are no longer organic and descriptive but are now abstract. "Pure" form. Formal form. It is no longer, according to Kandinsky, a space on a surface for sketching out a material object: form is a tension that exists *by itself*, a being that exists, acts, and exercises an influence. "It ceases to represent in order to become representational or representation itself."[9]

To liberate form from the object, Kandinsky purifies its representation as a bare layout.

Forms are actualized in their bare aspect, excluding any temporal or historical dimension. Forms, according to Kandinsky, establish themselves as a veritable code by means of a univocal correspondence between pictorial signs and "elementary instances of affectivity." Thus we find a stable system of equivalence; for example, Vertical—White—Activity—Birth. Thus, a catalogue of the forms of absolute consciousness is produced in Kandinsky's works. "Modern man seeks an inner repose because he is deafened by the outer world, and he believes the repose rests in inner silence …"

This is what abstract art furnishes. It moves in two currents: "A new objectivism in abstract compositions, which seeks in color relationships the production of objects that are purely formal rather than realistic." And then, on the other side, informal art, which seemingly rejects objectivism and which gives priority to matter and the primordial element by a total rejection of creating forms. This constitutes, in effect, not a basic movement toward aesthetic investigation but, rather, toward the true reality of the technical world. Now, here we confront an agonizing question, which shows to what degree art is in crisis: *technique*, as understood by Combet, abrogates all structures, forms, and differences. Combet demonstrates that technical production takes as its starting point "unformed, hardened substance" obtained from the processing of an original material to produce materials ready for decoration. To give them form they must be colored, varnished, lacquered, oxidized, etc. The outer appearance of the material becomes a pigment, or a resurfacing applied to nothingness. Art is decoration, the end we have inevitably achieved! Combet repeatedly shows that the mechanisms of *technique* tend to disappear.[10] "The function of *technique* is unreadable and remains invisible. The container has no relation to the contained." It becomes an all-purpose box, which tends to diminish in volume with miniaturization. "Minimalization of form leads to its complete disappearance."

"The function of the object devours all; form becomes useless. There is no longer any form at all." "The aesthetic of resurfacing and of enveloping, marginalization of forms, all purpose boxes … what has happened to functional forms? What of organic form? … Are we heading for the aesthetic of the gear box? It seems that the irreversible progress of *technique* leads to … the regressive evolution of forms … as witness to the development of an inescapable principle … the principle of the degradation of forms." All of this, deeply rooted in experience, shows the extent to which current aesthetic formalism is a gratuitous formalism with no reference either to man or to the world. Art is removed by *technique* from the universe of form, but, at this very moment, the artist, having abandoned meaning, decides to be only a creator of form. Under these circumstances, the term formalism must be challenged, but what will replace it? And these creators of form, having banished meaning, cling nonetheless to an aspect of meaning. Must one like Albérès, in his *Littérature Horizon 2000* (1974), connect the themes of the novel with the "milieu" and notice that character, plot, and author in the novel have been replaced by the city and linked to the theme of anonymity. But then where is the technical environment? Novelists are reactionary on this point. But then *technique* expresses itself through them unbeknownst to them. The obsession with discontinuity replaces the clear analysis of the classic novel, and relativity supplants the former stability of reason in the completed work. Or, on the other side, must one follow the example of the magazine *Change* (1974) and locate schools of thought in relation to intellectual movements: surrealism with psychoanalysis, literature committed to phenomenology, and the *nouveau roman* with structuralism. Here we are plunged into idealism. But, at any rate, form is the only way out for the novelist, the artist, or the creator, in order to escape the impasse of closure in the imaginary museum or to express, again, the inexpressible, and the shocking, confusing distortions of our civilization that are formal and technical at once. Without meaning or a sense of being, only form remains. Perhaps it would be useful to return to Aristotle's theory of form or to the theology of form in St. Thomas Aquinas to understand the importance of the phenomenon. Our project is less ambitious. I would like to see a challenge to the opposition between form and meaning as was done in the *Cahiers internationaux de symbolisme* (1967). Forms are meaningful by themselves—memory and imagination find their true places in the *nouveau roman*; one can prove that the labyrinth is a meaningful space, etc. All that is clearly true, but the great innovation is precisely the creation of forms that reside in language.

Of course, it is always possible for the serious critic—for the patient hermeneut—to find in these forms a rich source of meaning. Meditation on the abyss and the labyrinth is always very fruitful for a spiritual intellectual. But forget about a relationship or a communication or, even less, about something taken from life. We have an example in Jean Thibaudeau. In his work *Voilà les morts, à notre tour d'en sortir* (1974), the author sets us adrift in a work that is chopped up arbitrarily. It is text of plurality, which destroys discourse systematically and which, thanks to the driving force of its material, seeks, "the most direct route to the unconscious" (the generic "one" of the text being most deeply identified with the subjective). But how can one not see a formalism in this use of white, deserted beaches, and of all possible typographic signs, upper case, italics, etc.? This is the same formalism one finds in Cortázar and Fickelson and one that falls precisely under the critical gaze of Debord and Charbonneau, different as they are. For Debord, we witness a congealed consummation of spectacle with a former culture. The *nouveau roman,* as with all modern formalism, seeks to communicate the incommunicable in complete accordance with the dominant state of things in which all communication is proclaimed impossible despite its proliferation. This destruction of meaning and language hides the fact that our society is a society from which meaning has disappeared (and it pretends that this is not the case). The school of new literature, which makes the written word an object for contemplation, is precisely what allows us to avoid our world's plaguing problems. How better to characterize formalism? We see only what we write! And without entering the debate on structuralism, we limit ourselves to Debord: "The illusion that there is an unconscious pre-existing structure that dictates social practices may have mistakenly been drawn from structuralism and ethnology … simply because a universal thought is in an average setting … a thought totally rooted in the overwhelming approval of the existing system … confers reality on that system … Structure is the daughter of current power. Structuralism (and here I would say the structuralism of *modern art*—Jacques Ellul) is thought guaranteed by the state, which ordains the current conditions of communication and spectacle as an absolute. Its approach to studying the code of messages is nothing but the result and the recognition of a society where communication exists under the form of a cascade of hierarchical symbols. Thus, it is not structuralism that proves the transhistoric validity of the society of spectacle: on the contrary it is the society of spectacle imposing itself as overpowering reality which proves the cold dream of structuralism." I have nothing to add by way of demonstration to the subject of this formalism in fields of art, painting, sculpture, and the novel. All are simply pieces in the society of spectacle.

Charbonneau and Mumford predicted this evolution much earlier. The novel, for Charbonneau, is condemned to disappear on its own power, because it is "the expression of the individual who creates society by becoming conscious of its presence." This conflict is central to the classic novel in which a society of individuals is presupposed. With the advent of totalitarian society the novel disappears. How does one continue to express a situation where there is no relation with others, with taking on the world, with no faith in God? "For a man thus reduced to himself, what is there to do? For lack of anything better one talks to the world about oneself, or one maintains that there is nothing left to say. With nothing more to say there is always literature." (The essential point is to remember the identical nature of talking to the world about oneself and there being nothing more to say. This is precisely what modern art does: pure subjectivism is contrasted with the all or nothing.) For "when the novelist is devoted to a photographic description of behaviors, he reflects a society which considers men as things …" And Charbonneau entitles this entire chapter, "Speaking to Say Nothing." This indeed characterizes the *nouveau roman,* and this should bring despair, but it does not. Despair is still an expression of humanity. There is no longer despair because man disappears in a cold flood of glacial break-up. Now we come to a central point, which Mumford has already revealed. In the first place, the formalists of the *nouveau roman* have the pure technical mentality: one acts on language, on writing, on the means, and when one is done, one acts on the whole, i.e. nothing else remains. No beyond of any kind appears beyond the instrument. Every current technician experiences this. Furthermore, it is interesting to find in writers the same psychological attitude that we see in all technicians and technocrats: contempt for people who do not understand their techniques and no sympathy for those attached to ancient values and forms. The technician strives to always go beyond to achieve efficiency, to refuse to explain to anyone not on his level, to grasp an absolute truth in the practice of his art. However, the relationship of *technique* to the public is a work of assimilating and reducing the public to the values of *technique*. The *nouveau roman* functions principally for intellectuals by accustoming them to see nothing else but a play of means; i.e., by imprisoning them in the technical system. This formalist and theoretical art plays a double contradictory role: it professes to revolt against our hyper-mechanized, hyper-regimented culture, but actually justifies the products of the system in power. It acclimatizes man to live in these cities and in this milieu; it convinces him that this absurd world of violence and anonymity is the only possible world. It makes him consider his life in large housing projects as normal and as representative of the pinnacle of art rather than as symptoms of his disintegration among such products of mega-technology. Protesting against such an environment seems absurd. Rather, this milieu forces man to accept as his true being what is in fact its negation. It makes him accept as natural and good an environment degraded by human and industrial waste, by nuclear plants, by super highways, by pollution, all of which destroy the natural environment. Sculptures made from pieces of machinery or from crushed cars or paintings of a nuclear desert bereft of nature are not artistic confrontations with the modern world but are rather a means of conditioning the hypnotized and desensitized masses to their situation. "See here, discarded waste is not so bad, since art itself finds inspiration in it." They justify the civilization of discarded waste. The artist increasingly integrates what is human into technical acceptance. Thus one can refer to Hans Bellmer and his mechanized doll with its ball-bearing joints, which is intended to express and provoke eroticism and is sexuality exacerbated by its inhumanity. J. Bousquet wrote, concerning this, "all that is real is converted into all that is imaginable," but this is a real created by *technique*, which, conveniently manipulated by the artist, incites the erotic imagination; it dehumanizes man by leading him to take a mechanical doll as his partner.

Modern art (we have seen numerous examples of this) is characterized by a destructive intention, not only of art itself *in abstracto*, but also of art works of the past. Thus, we have theatrical revisions of the classics and of opera (*Parsifal* or *The Magic Flute*). They undertake to deconstruct, to unravel, to disorganize, and to attack meaning, but for what purpose? If I go beyond the current banalities (modernize art of the past, or reinterpret it in terms of class struggle, or abandon bourgeois culture), I can find only one interpretation for this destructive enterprise, which exists only through what it destroys: the aim to remove any obstacle or fundamental challenge to the dominant force of our society, namely *technique*. The function of cultural modernization in this art is the work of integrating man into the technical universe. The Dadaist poets, the authors of the *nouveau roman*, the atonal musicians, the painters and sculptors of theoretical art, are the handmaidens of the technical system. There is nothing in them, and I emphasize *nothing*, that is worth retaining or considering. They are part and parcel of the conditioning mechanisms engendered by the technical system. Their ultimate betrayal is to add the sugar coating of freedom to this radical degradation. The lie of modern art is in its claim to be the perpetual expression of freedom. These artists who claim to be minimalists, serialists, structuralists, and objectivists convert their subjective nihilism with the system into an object of delight, artistic creativity, and inspiration, and by so doing, they support the illusion of defeating destiny through an act of personal choice. Not only is there no choice, but there is only an abject acceptance of results programmed in advance that betrays man's last line of defense. They are the Trojan horse that introduces doubt and corruption and that intensifies the rampant irrationality underlying the system of technical power. Here, we witness a remarkable fact: the technical system, which is essentially rational, produces burgeoning irrationalities; its rationality and its irrationality go hand in hand.[11] Furthermore, we find the exact duplication of this process in theoretical art; the more theory increases, the more the production appears irrational, incomprehensible, and a-rational. The more the concept is mathematical and systematized, the more the consequences are inaccessible and unsatisfying. Once again we find a relation between these two aspects, both of which are the characteristics of a mechanism of power that eliminates in its development all that could connect the creator and the spectator to traditions, values, and rhythms that are indispensable for reconnecting and reorienting man to this universe of forms and means. To the degree that new means are exclusively oriented toward the new and have a power such that they can, for the first time, be all that constituted human experience, they can only increase the growing distortion between rationality and irrationality. Now, this growth of irrationality will deliver man, through the intermediary of art, either to a superficial delirium (which we see in cold eroticism or cool music) or to a system of absence where nobody talks to anyone. Things are done and that's what is talked about, and with this fact, the individual, if he still exists, finds himself increasingly adrift in a meaningless existence, which causes the most profound distress. And this leads to another function of formalism. What is the root cause of purely formal art? In reality, art is no longer able to symbolize anything whatever.

We have seen why symbolization has become impossible in the technical milieu. However, art does not exist without symbolization. Modern artists have finessed this aridity thanks to the creation of forms. These forms maintain the appearance of art but only to the degree that they have no reference beyond themselves. Artists have taken the step of removing meaning as a necessary precondition for the absence of symbolization—and once again, these strutting little cocks, mounted on their high horses, crowing at every echo that comes their way, are buried under an avalanche of absurdities. There is no symbolization, and form remains a simple sign of absence.[12]

This absence is a signpost that indicates authoritatively a road toward a city, which has disappeared as in an atomic blast. What's to be done? Carefully sculpt the post, paint and repaint imaginatively the sign itself … while affirming this is the answer. Formalism increases to the extent that symbolization is no longer possible. Formalism, turning back on itself and exhausting itself, with a sterility that does not nourish, is a substitute for symbolism[13] nostalgically sought, as necessary for man as bread and wine—hence the creative rage of these formalists with the rush of their productions. They pile up thousands of canvases, poems, films, and symphonies that produce, at best, one symbol or, perhaps in their cacophony, a silence in the absence of symbols. One speaks loud and strong with the hope of finding a place in this vacuous blackness of interstellar space that we can travel but not possess, for only symbols can truly master it.

Because formalism occupies this vacuum and accentuates and *legitimates* the impossibility of symbolization, it has but one final but important consequence: for millennia painting was a means of defining "reality."[14] All arts partake of this relation to human reality. Now, however, the theoretical art of our time fulfills precisely the opposite function. Thanks to these methods reality is fractured; the individual's purchase on reality, already weakened, is rendered impossible with the substitution of a wild subjectivism that adheres to the social processes.[15] Formalism detaches the individual from any human reality and leads the spectator to the same detachment from what had constituted, until now, the human reality of a milieu in which man lived and moved and had his being. This detachment appears precisely when the technical milieu comes to dominate as a reality with which man finds himself at odds and to which he becomes a perfect stranger. Art embraces this new *reality*, but in so doing, shifts to the other side and no longer provides a means for man to grasp human reality. This new reality, although created by the human, is ultimately foreign and inhuman. And since this art is the product of hermetic and esoteric specialists, the artist no longer plays a traditional role of interpreter, intermediary, or mediator. The supreme artificiality and sublimity of this art deprives the artist of his origins and justifications.

All of this art is so perfectly artificial in the *sense of false* that one can claim in each instance that no more can be achieved, which in effect means the end of this type of art. After the famous white square painted on a white canvas by Kasimir Malevich in 1918, one could conclude that the end had been reached. After Victor Vasarely, Malraux declared, "painting is finished." After the *Anti-poème* of Denis Roche, one could write that, "Henceforth it will be no longer possible to write poems." But, pardon me, things do continue. Things continue in a delirious attempt to outdo the past. Since nothing signifies nothing, all is possible—a veritable circus with more and more sideshows that prove without doubt that pure formalism, grounded in the mechanisms of technical performance, has replaced the former schools of art. Even the spirit is devoured by *technique*. One last comment, perhaps totally useless: the artists of whom we have written and who are so heavily in the grip of the technical system in doctrine and theory are not the "co-opted ones." We have seen in the preceding chapter that revolutionary artists with a message are habitually co-opted. Here, this is not the case. These artists are not exploited against the will they do not have by the nasty capitalist bourgeoisie. They are already marvelously in tune as the pure products of *technique* and are completely at the service of the system, pure and simple. They are not the willing tools of the capitalist world but of the technical world, which knows how to use and exploit them. They may call themselves revolutionary, but that's irrelevant.

There remains one last point that reveals one last contradiction—the relation of this art with the public. Here we come to a remarkable imbroglio: faced with this modern art the average man is indifferent. In front of a Vasarely, the viewer casts a passing glance and moves on as if it were some piece of wallpaper or linoleum that is of interest to him only if he is redoing his bathroom. The listener regards modern music as background noise easily ignored—it could be swing, it could be Bach, it's all the same. That is to say, this "art" is perceived essentially as a form of decoration. (And, indeed, the "paintings" produced by computer fully deserve to be considered wallpaper). If one explains to the average person that this is Art, great Art, he shrugs his shoulders at this joke. At most he can be interested for five minutes upon being told that such and such an artist paints with a blunderbuss and that another smears his body with paint and rolls on a canvas on the floor. It is sickening. If one attempts a justifying discourse on the metaphysical foundations of this art, he will consider these artists a bunch of loonies. At a slightly more elevated level, Jean Bloch-Michel (*Le présent de l'indicatif: Essai sur le nouveau roman*, 1973) in his criticism of the *nouveau roman*, explains perfectly the reactions, the critiques, and the rejections of the average reading public. Boredom. In the *nouveau roman* there's an absence of novelty, to the extent that anything is comprehensible—there's preciousness, grandstanding, and posturing, with a lack of creativity and inspiration. As for usage, the present indicative of the verb is interesting in that it allows for the description of a world without *true* existence. The cultivated person is in effect aware of the false character of this music, of this painting, and of this kind of novel. Now, here is the paradox: theoretical art cannot survive without public support, which is the sole criterion of its validity. Indeed, there is no longer any guarantee of authenticity or any reference either to classical thought or to a faith in truth as in the case of art inspired by Christianity or by a *sensus communis* as in Greek art. Nothing. The theory pushed forward does not legitimize or ground this new art in past precedent. There is only an outreach to the public and to the backing of a fragmented and uncertain crowd to be won over. Only in this way can this art be validated, which often requires the inclusion of fervent declarations against elite and privileged art. Only a general plebiscite, as Moles says, can define the work of art. "If the world is full of beautiful things, the critic will be transformed into an artist as soon as he puts a framework around a random piece of macadam with an infallible gaze that assigns aesthetic value. It is noteworthy that no one is responsible in this business. This approach to art is based on the *consensus omnium* of mankind." (The critic is the mouthpiece of this *consensus omnium*. But, did Moles ask himself what the untutored public thinks when it stands before a framed piece of macadam?) "The definition of the Beautiful is based on statistics concerning the Beautiful." (Indeed, it is essential to remember that, for these theoreticians, the Beautiful is what the majority thinks it is.) One must "take an interest in works of art without judging them either good or bad except on the basis of a social sense." Only the backing of the public makes a work beautiful and, hence, worthy of being considered art. Daix or Delevoy also cite indirectly success with the public as a criterion. Manet, the impressionists, and Cézanne are great artists because the public (which public—the art dealers, the specialists of exhibitions, or "the public in general" … the *consensus omnium*?) has embraced their vision. Enough said. These examples are compelling. They lead us to two trains of thought. First, the chosen criterion is not, contrary to appearance, the democratic criterion! It is the criterion of *technique* that validates success. *Technique* judges only on the basis of efficiency. Therefore, in an area like art, no other criterion of efficiency exists beyond the backing of the spectator, of the masses, of numbers, since these forms of art are disseminated by the mass media. To refer to a plebiscite in order to define the beautiful, far from being a democratic idea, is rather a technical attitude. But this leads us to a dilemma: must one consider the public as it currently exists, as we know it, or as the public in general in its immediate reality? But we know its reaction in the presence of modern art. Therefore, we should stick to Millet's *Angelus* or *New Orleans* or else, if we are convinced that modern art is truly important, rich, and valuable, (but here we abandon our own criterion!) then we must educate the public to appreciate it. The public must be made to like what "one" has decided they should like. We have referred to Durailh's interesting declaration at the Festival de Rouen (1975), according to which there is a prevailing "misunderstanding" concerning modern music: one judges it according to a traditional ear, which is inadequate. One must, therefore, educate the taste, the eye, the ear of this public in order to gain its backing. We have already encountered this contradiction. Therefore, we are facing a new type of technical attitude. With each technical advance there is a reluctance by the user (pilots of propeller planes initially refused to pilot jet planes) and with the advent of every new product there is a resistance from the consumer. From the technical perspective all new products are good, *a priori*, because they are new; therefore, one must mold man, convince him, acclimatize him, indoctrinate him, and manipulate him until he utilizes and consumes the technical product. This is the task of propaganda or advertising, of education at school, or of upbringing. One manages to persuade the reader of Cortázar or Roubaud that reading is basically a game with which he is able to spend pleasant hours solving puzzles or that watching television is like a game of Go. If this happens, *technique* has won. But, one must manufacture a public and lead it to adopt a product according to the false criterion I have referred to, and thus there is no *consensus omnium* except after the fact. The only criterion is that of the new and that of the potential of the device to win over the masses. The public's curiosity is attracted by a nifty trick (*Les Nanas* of Saint Phalle), by apparent audacity (pornographic film), by institutionalization (the festivals), by the sociological phenomenon of creating an in-group (those who have been at Avignon and those who have not been, or, a thousand times better, those who have been to Shiraz for the Bob Wilson festival), that is to say, by processes of setting up elites and of putting them in motion through techniques of mass manipulation. Under these conditions, modern theoretical art finds an audience. By itself, it is totally foreign to this excluded public. Furthermore, it seems that the modern artist wants to exclude them at all costs. When Jacques Rivette offers us the film *Out 1*, running twelve hours and forty minutes, one understands how little this would appeal to the public. One can speculate that Rivette creates new relations between the spectator and the screen that are immersions in cinema, which are calls for original reflection in that the spectators, during the breaks every two hours, are urged to exchange views about a film that is limited to projecting to the audience (without needing to dramatize) the life invented by actors without roles, characters in life based on a "plot" without any consistency that does not represent even a hint of story-line. All this is fine and good, very new, but who are the spectators with twelve hours and forty minutes to spend on a film? Who, intellectually, can tolerate such a test? Who can put up with a complete absence of coherence? Indeed, we are witnessing an enterprise of aesthetes for aesthetes, of specialists for specialists, work totally dedicated to in-groups. And it's exactly the same thing when Ricardou shows how to read a modern text, especially one that is a *nouveau roman*. From one hundred examples he shows that one must collate sentences taken from one novel with those of another by the same author to discover the true meaning (a similar narration): in other words, one must have an encyclopedic background, a prodigious memory, (and better yet, a filing cabinet of the novels one has read). The sorry fool of a reader who reads this story after a day at work and who gets vaguely interested not only does not understand what he reads—and sometimes he realizes this and abandons his reading in discouragement—but still does not even find himself on the threshold of the holy of holies. Moreover, there is an obvious condescension for the non-specialist. "For the naïve reader who engrosses himself in the most essential aspect of a story, such a reassembling of its parts is close to sacrilege."[16] (Ricardou). Sacrilege is saying a lot but it is clear that this practice takes away all pleasure from reading if one must endlessly pose enigmas and search for correspondences as in the true work of the alchemist in the 1600s or a Kabbalist of the 1300s. Casually, Ricardou poses the question: would this not be elitist? "An in-the-know pessimist would claim that any active reading could only be a privilege of caste reserved for writers. A revolutionary optimist would read in this, rather, an incitation or a call to all to practice the work of writing necessary for all reading." Take note of this slippage: anyone who claims that to do such difficult and recondite analyses to work must be nicely stuffed with facts, and he is the true pessimist in the know! But, one wonders why the man in the street doesn't get his kicks out of Roussel[17] and why these literary games are played by a very tiny team. Ricardou's reflection expresses, in reality, a great blindness that one sees in the technician who believes that his technique (so simple, so evident, so rational, so it is said) is accessible to all. In reality, the public is effectively excluded from this art of Mandarins, who depend, as we have shown, on the support of the public and the *consensus omnium*. But here, in effect, without their understanding, the public is passively absorbed and, little by little, taken in by theoretical art devoid of meaning. They accept it (this is very clear in the style of popular publications and those of films for the past ten years). They go along because there's no alternative. To lead the public to accept meaningless works is possible because they live a life without meaning, which only aggravates the meaninglessness of life. Highly artistic formalism accentuates and completes the work of television in spite of its declared intent, which is immediately co-opted by its premises of involving and leading the spectator himself to write and to play. Led to play, to create? At once the average man will ask the question: "Yes, but why?" What motivation would he have to be involved in this effort? Because it has no meaning, there is no motivation besides recognizing it for what it is, and if there's no motivation, there's no chance for the public's involvement. In other words, theoretical art only reinforces passivity, the attitude of the consumer, the Mithridatic road to indifference. This indifference, carried to an extreme, goes hand in hand with the proclaimed absence of meaning.

Notes

1 Each art exhibition requires a catalogue containing a long introduction by a specialist or an interview with the artist in question, without which no one can penetrate the obscure and recondite world of the exhibition. Discourse is needed to complete and explicate naïve sensibility and understanding! A good example of this is the catalogue for the exhibition *Identity-Identification* which belies its title: non-identity without identification would be a better title.

2 Anne Cauquelin signals, similarly, the important combination of the painter and the philosopher: "The painter needs the verbal support of 'his' philosopher, and the philosopher who has lost his own raw material in the flood of specialization must expatiate on an object of art, preferably pictorial art. ... Thus we have such combinations as Monory-Lyotard, Premenger-Deleuze, Manager-Lascoubet ..."

3 Here we find a juncture with revolutionary art. Tàpies is a good example. One can decide to make a revolution in a pot of paint, but in a way so obscure, symbolic, abstract, remote, that no one understands it. Painting in red and black—blood and death, so be it—but what revolutionary message is found in collages of wood and straw? One would have had to have taken the road of Tàpies himself and grasped the symbolic form of everything by proceeding through an association of ideas, which is like saying nothing. In other words, technical rigor proposes to empty all forms of art of meaning.

4 cf. *The Technological Society* and *The Technical System*.

5 Ricardou, *Problèmes du nouveau roman*, 1967; *Pour une théorie du nouveau roman*, 1971.

6 One can thus classify Rivette's enormous film *Out 1* as a "puzzle," made of unrelated fragments of realist cinema, which capture the innovations of the "actors," and, thereby, each viewer must assemble for himself his own film following such and such a plot line; the viewer must find a key to the film beyond the film and attempt to respond to the questions that the work poses without providing any hint for the answer. It is exactly the same in music; how can we not consider the *Archipels* of Boucourechliev (1968) as anything but a game, whose elements are moveable and can be arranged by the performer in infinitely variable ways that lack any theme. When the variations have radically undermined the theme, then there's nothing left but play.

7 Ellul is giving an intentional misspelling of "Voulez-vous jouer avec moi?" "Would you like to play with me?" [Trans.]

8 These innumerable puns have no other function than to trick the reader out of an expected meaning. We see this in *Le voyage à Naucratis*, by Almira, (1975), 550 pages of verbal delirium with no other clue to meaning beyond the author's Me, Me, Me, *ad nauseam*. And extracts from the magazine *Tel Quel* featuring the cosmogony of Brisset are a series of puns on teeth and mouth. [again impossible to translate]. For the interpretation of texts, we have the same subtlety: Ricardou establishes the correlation within a novel where the word has taken on diverse forms. [Again, cannot be translated]. We are in the presence of the destruction of language when it becomes a pretext for juggling.

9 Concerning this we have followed Maldiney's study, *Regard Parole Espace* (1973).

10 I disagree with Combet and his wish to reintroduce the artistic sense into the body of our industrial society, but I agree with Combet that art has become an aesthetic of the invisible.

11 On the relation between technical rationality and technical irrationality, cf. Bernard Charbonneau, *Le système et le chaos*, 1973; and Jacques Ellul, *The Technological System*.

12 Abstraction ends by falling into absence. We see this in American postminimalism. (*Exposition Arc 2, Musée d'Art moderne*, 1975). Two charcoal lines, scotch taped, fastened to a piece of paper, two lines forming an X, a black rectangle inside a white rectangle ... reduction to the limit, clearly and radically opposed to all humanism and all thought. Purely geometrical forms correspond exactly to a technical environment (or even more so).

13 There is no point in asking the reader to give me credit for the term symbolism, because in no way do I intend the artistic, poetic, and painterly *school* that was in the nineteenth century symbolist.

14 Psychiatrists have also discovered that painting was one of the means of allowing a patient to find a road back to reality.

15 When I speak of adhesion and conformity to the processes of society, it is obvious to me that one can be a Maoist, a leftist, a nihilist, etc, while being a perfect conformist to present day social life for which political options have no meaning.

16 This means the parsing of the text for purposes of filing, comparison, and typology.

17 Of course, I know the response that may be raised to this objection: the proletarian does not involve himself in the joys or Rousselian exegesis because he is stupefied and reified by the dominant ideology. I do not reply that I don't see any so-called socialist state where the public could join this game. Likewise, it seems that outfits like the Festival d'Avignon are only possible in a liberal capitalist regime, and that the exercises like those of Ricardou would be un-publishable in China. This is much too exaggerated. But I am still waiting for them to explain under what dispensation—concrete, not some schematized paradise—everyone could get involved in such abstract creativity, unconnected at the same time that it is bereft of meaning.

Chapter 5

The Artist and the Critic

I / The Artist[1]

What is the essential nature of the artist in this world? Who is this creator of a tattered art of prophetic exaltations and of sibylline enigmas? What is his image, his ideal description, on the one hand, and his effective place and role in the technical system, on the other hand? Of particular interest here is the encounter of a bloated discourse in its pure state. It is a discourse conducted by the most learned intellectuals of our time that has no relation to any reality except the reality of discourse. Once again, we find that art occupies an exemplary place in the entire technical system that we have been examining to discern the relation of artistic discourse to the technological society that has become a system.

We begin with the claim that the artist is an absolutely free individual, and thanks to *technique*, he is much freer today than ever. It is clear that the artist in the technological society enjoys an infinitely greater freedom when he knows how to utilize the means of *technique*. He has a freedom, for example, in the exercise of means, which applies to the painter, the sculptor, and the architect who possess technical means of which their predecessors could only dream. The artist can invent freely without concern for its realization. New techniques give the artist an extraordinary mastery of material without the use of tricks or indirect means to overcome the obstacles of stone or weight: the machine liberates both the sculptor and architect from all impossibility. And *technique* provides new materials. If he cannot bend traditional materials to his will, he can make use of artificial materials, which, since they are isomorphic, neutral, and unlimited, will perfectly serve his purpose. Thus, the artist finds himself completely free to conceive.[2] He knows that the means make him master of the production, however audacious, unforeseen, or extravagant his ideas may be. Henceforth art is no longer tied to a predetermined way of using the traditional materials. There is a double transformation.[3] First of all, the artist receives from *technique* what one can call an active concept of the material: that is to say, that the material itself becomes an element of art. The material *per se* is a work of art that coexists with an emptiness to which the work of art confers value whatever the artist's intention. The architect liberates himself from the concept of a wall to create a building where the communication between an exterior and interior space becomes important and will be enlivened by light. Moreover, this is the moment when the sculptor empties solid form and ends up with the simple line of a steel wire. The other transformation in the technical world rests on the fact that the artist can create his own material. He is no longer tied to material limitations that hamper his creativity: from the technician he obtains an assembly of powers that allow him to freely act. Thus, through this double transformation, he no longer knows external limits. His creative activity is entirely up to him. He can create what he wishes, thinks, and invents, and he knows that in all cases this creativity is feasible. Henceforth, the artist "knows" that art relies on forms becoming signs that construct a world in the image of the imaginary and no longer in the image of the real. The art of creative signs that fade away and replace the direct dialogue between the spectator and the creator encompasses "man himself, his history, his place in the world, his freedom, and the very resistance of nature to his understanding." (Roland Barthes). From this point on, visual or sonorous forms depend upon the freedom and control of the artist with theory. And theory allows for marking out and even exploring the entire field of possibilities which open up and which are the very freedoms of the creator. To the degree that all relation to the real and to all previously existing art is severed, the artist is truly free to do what he wishes and to engage his pure fantasy to assemble any object of sight and sound. New materials and new instruments open all the doors that lead to this fantasy, which, in turn, leads to the proliferation of schools, approaches, and failures. Periodically, specialists proclaim this freedom. J. Michel, referring to street painting and subculture, would write: "*La liberté conquise?*" (*Le Monde,* January 1970), in agreement with Malraux, stating in 1959, "Painting has discovered its freedom," with Tinguely being the epitome of this freedom. Of course, the artist intends to enlist the spectator in this freedom: the architecture of Beaubourg seeks to create a machine for communication and invention. The artist resigns as a discoverer of forms and means of expression, and also of materials and means. Like a veritable chemical engineer or physician the artist creates the material he needs. Everything yields to the force of his freedom. He is limited by nothing, by no constraint or precedent, or no imperative. He repudiates form and meaning while acquiring autonomy in the material realm. This opens up an area of infinite fluidity (which gives, in the artist's words, substance and meaning to the work). "Informal art is the first historically complex style in which no element is predetermined, where the sign precedes signification and postulates its ambiguity." Informal art is absolutely creative because it depends on nothing. And we defend this situation, relying on Hegel's affirmation, which announced (already, but a little too early) the death of art, by stating that the artist exists even if art is dead. "The artist who embodies fantasy and freedom is the safeguard of mankind." (Leymarie). Furthermore, the artist moves to the margins of society by challenging it in all its aspects and, in so doing, gains his freedom. Of course, one can wonder, with a little anxiety, why such freedom neither produces nor provokes any outrage, any conflict, nor any opposition. This may be the first time in social history that an authentic freedom does not cause outrage. We will come back to this. Georges Mathieu expresses this freedom by exhibiting a formula of lyrical abstraction: his painting is abstract when it is no longer based on any subject and is lyrical because it opposes all that is concrete, fixed, and anticipated. Thus, freedom has two sides. One is no longer obliged to follow a model but is able to challenge any style, any school, or any institution. Freedom becomes the ability to create and to bring forth new potentialities, and, thus, a new culture is created. We experience the origin of creation, "the apotheosis of risk, the festival of being." The artist creates in a void (we know that if placed in a bell jar with a perfect vacuum, one would enjoy incredible freedom). This is the amazing smugness of the artist who is absolutely unaware of his servitude and of his limits; his intellectual faculties are one hundred years old. And when Mathieu wants to declare his absolute independence he makes posters for Air France and tourism and creates postage stamps. Unconsciously, the artist creates consumer objects for technical society.

Likewise, this freedom of "whatever or anything goes" is well revealed in Cage's work: "Each being and, likewise, every sound, can be considered as the center of the universe. There is no single focus—the work—but a multitude of centers that must be respected. In order to play a score like this one, everybody must be perfectly free." On this basis, one can do anything. A work like *Atlas Eclipticalis*, 1962, is written for one to ninety-eight orchestral musicians with twenty five thousand sounds freely played. One could just as well play such works simultaneously or separately. At the Festival of La Rochelle in 1976, *Atlas, Winter Music,* and *Solo for Voice 45* were played simultaneously. Furthermore, Cage takes a literary text and proceeds to mix up syllables and letters according to the operations of chance, carried out thanks to a collection of oracles from ancient China. That's called freedom! Freedom carried to the extreme! That's all we can say in the face of this contempt, of this incoherence, of this non-being produced by a total absence of understanding of our reality, despite Cage's claim of concern for society and politics. This is simply a contribution to the disintegration of art and to the clash between Art and *Technique*. Nothing more.

The situation is very clear: all is possible, all is permitted, and all can be done intellectually, sociologically, and materially. Abstraction undermines the rules and leads to perfect absence. To do everything is the triumph of "whatever." The line the artist draws is the real itself. The line takes on meaning simply because it is real; it was meaningless but now becomes real. "The glory of whoever and indeed of whatever simply belongs to that which is and reveals the glory of painting." (Georges Bataille). But this concept of "anything goes" is chaos. For happy spirits this chaos is salutary and blissful. (Thomas B. Hess). Only the partisans of coherence and meaning complain. It is clear that the public itself needs this "whatever."

It is quite evident that when one is brought up from kindergarten to understand the music of Boulez as music and to allow the painting of Tal-Coat as an aspect of painting, then one knows instinctively that painting or music can be anything. One can torture colors and sounds in any way whatever, but then it becomes necessary (since one has eliminated in advance all criteria of beauty and pleasure) to indoctrinate the public, and the work is done. But I would like, here, to advance an impolite argument. Until now we have written only of the "Great" Arts. We should not, however, overlook the art of cuisine, which, since it affects one of the senses, should be taken seriously. Cuisine is the basis of the social group, and the meal is one of the fundamental elements of community. Cooking is a basic art of culture. It is unnecessary to have read Lévi-Strauss to understand this, although he has masterfully demonstrated its importance. Cuisine is an art rooted deeply in the cultural world of the people where everyone exercises a high degree of aesthetic discrimination. Culinary art, however, has disappeared. We no longer recognize the great successes of traditional cooking, and whether one likes it or not, the industrialization of canned goods reduces the capacity for taste in strange ways. The very raw materials of cooking have been falsified (for example by the meat of animals raised in mass production). This should give pause to those who become enthusiastic about freedom in regard to the materials of art, but let's return to the idea that everything is permitted. Try this in the kitchen. Consider, for example, the following recipe: take a glass of paradichlorobenzene and add to it a large tube of neoprene glue, throw in a trickle of ascorbic acid to bring out the flavor. Reduce on a low flame. Then cut some expanded polystyrene into wide slices, dip them in a sauce made from oil drained from a car, and serve hot. You can torture the ear and attack the eye, but you cannot force taste to accept anything or "whatever." Such is the limit of reality. But, the more you approach the true, the more you reach the centers of abstraction, of reflection, of mediation that is mental, the more the possibility of lying, of doing harm, or of destroying increases. Truth and "spirit" do not defend themselves like a stomach in revolt, and this is the problem. One can break apart language. Clearly meaning and communication do not produce the immediate reaction that one would have to sulfuric acid. Modern painting, music and literature—because they obey the law that "everything is permitted"—are in the same order as my cooking recipe, and since the effect is only on the nerves, on the psyche, on the intellect, on the ethical and spiritual make-up, none of which registers on a seismograph, then there's no need to worry. It's a simple matter of adaptation. The public is ignorant, and they must be educated, but to what end? But this is all the more harmful, all the more contemptible. Today, could a single artist escape this? No, not a single one. And what criterion would he have for escaping? All has been eliminated in advance. From the notion that "all is possible and all is permitted," the artists of our time have entered not into freedom but into a situation worthy of absolute contempt, because they have only contempt for the listener and the spectator under the guise of communication, participation. The problem is not participation *per se* but rather the problem of participation in what, the communication of what? Without participation in something and without communication of something, there is only contempt. But there is more. Does liberty exist? Some will doubt it. Mies van der Rohe said, along with many others: "It is not good to have too much freedom; constraints are the best stimulants of the architect." And if the real, if tradition, if meaning no longer impose any constraint, then the artist will manufacture his own little set of rules. If an exterior reference no longer applies, then one is supplied by an elaborate theoretical network woven out of thin air. We will see later on to what degree the modern artist is determined, to what degree his freedom is fiction and unfounded, and to what extent this freedom is merely a proclamation against an imaginary barrier that justifies and conceals the true constraints that are heavier and more tragic.

At first glance, it is surprising that we confront a freedom the artist cannot escape and which places him in a difficult situation, because the artist has never before encountered similar problems. And going back to what we wrote previously, the first problem comes precisely from the loss of traditions. In former times, the artist was embedded in a social and ethical network that led directly to a certain aesthetic style. He obeyed a set of rules and forms, which, simultaneously, limited but also extended his possibilities and genius to a precise destination. Tradition provided him with a stepping stone to go beyond; rules served to test the powers of his creative ability. Tradition proved both obstacle and springboard. Technical civilization has shattered traditional ethics and social control along with slowly evolving aesthetic norms. We may say that the artist has been thrown into the freedom of a void. The artist can do anything or everything but without having a base in tradition, he cannot avoid the anguish of an absence of continuity. He is condemned to start again from ground zero, because all has been challenged by the shock of techniques. It's not by accident that if one returns to the primitive arts to rediscover the value of most ancient forms, one will learn from those who also had to start from scratch, who viewed the world without preconceptions and who sought at this elementary stage an aesthetic expression just as the artist today is obliged to do. It is useful to realize that we are, indeed, in a comparable situation to primitive man, who struggled with a nature that was as yet unknown. Today we are grappling with a phenomenon both local and global that has replaced nature, which is the technical civilization to which we have not yet adapted.

One can go further: we have a freedom that turns back on itself and destroys itself. Adorno has clearly demonstrated what he calls "the fall back into non-freedom." "Music is an example of this dialectical reversal. The technique of the twelve-tone scale is truly its destiny, which enslaves music while liberating it. The musician subdues music through a rational system only to succumb to this system … Technique is manifest in the arrangement of material according to its criteria. It then becomes a determining factor, which, in its alienated state, opposes the musician and subjects him to its own constraints. Just when the composer's imagination has subjugated music to his own will, at that point music so subjugated paralyzes the composer's own imagination … The operations that shattered the blind despotism of sonorous material become, through a system of rules, blind second nature … The violence that mass music inflicts on man continues to affect those on the opposite end of the social scale; that is to say, in music, that goes over the average man's head …" Indeed, the attitude of "anything goes" engenders the response that "nothing matters." Absolute freedom means absolute subjugation to something as yet unknown. The drunken imagination simply wanders. "The kingdom of imagination belongs to a fallen king." Only a fallen king can revel in his imagination. Indeed this is precisely the condition of the modern artist! This proclaimed freedom is not the explosion of an unlimited force but is a social role: the artist is the specialist of freedom in a rationalized world. We must still have a possible and satisfying representation of the Individual with a capital I, a confirmation of his freedom.[4]

How sad this would be.

The world is harsh, closed in, unimaginative, ruled by *technique*, rigorous and hostile. We have neither leisure nor the possibility of self-expression nor the demonstration of freedom (which surely must exist), and so we place all our hopes on the specialists of freedom and personality. And the artist fulfills that self-imposed role. We cannot do without freedom or resign from being a person. The artist is a specialist in playing the role of freedom. This means that the artist is basically a play actor. All that the modern artist does is meant to call attention to himself and to his comings and goings and not to his work. And thus we see the incredible attention to staging in the theater or to exaggerated gesture to the detriment of the work, which must disappear behind the effects of stage-craft and actors instead of being served by them. This also relates to the innumerable interviews where we find the minute details of the lives of great modern artists. The reign of "Didyouseemeism" is absolute. The artist who, formerly, sought to create a work is now occupied with creating a personality, and that relates to what we have said about an intent that was especially notable in the generation of Beat Poets—to make an exemplary work of their own lives. And for a lack of anything better, this can be done through drugs and alcohol, and this is, truly, magnificently, a role to undertake. Ultimately (and this is often the case), it is no longer necessary to write a poem or to draw a line: it suffices merely to be there. The drugged artist, because he violates the awful bourgeois morality, is the great artist of his own life. A poet of life experience no longer creates a work but rather provides an act, a product, or an event. Certainly one can say that a happening on the *rue du Dragon* is in no way on the scale of that other happening which occurred at Hiroshima, but this thought is clearly idiotic. There is no comparison. It's the absolute of the instant, an event in its pure state, which produces a quick beam of light on the absolute and free personality of the artist. Likewise, this snapshot vanishes as rapidly as a flash bulb. This poses not just a change in style or a mind-numbing avalanche of information, which so radically eliminates the artist, but is his very concept of art. After the artist is gone nothing remains, especially not the artist who once held the foreground. On the whole, painters, novelists, poets are annihilated after their ascension and are forgotten when they leave the public eye. Musicians and cinematographers have a longer shelf life. As for architects, one is forced to endure their work long after they disappear from memory.

Whatever the case, the artist, burdened with a social function of embodying freedom, manifests a character beyond the norm and beyond public judgment. His destiny is unique so no one can understand or judge him. He claims to become the public conscience or at least its expression, but to that end, he must be seen, considered, and regarded as the author of ever new eruptions of creativity. Recently, a political or religious conversion is taken as a good career move. One becomes a communist (which is currently less chic) or an anti-communist after having been an unconditional supporter of the party (now that is more the style). But a conversion to Christianity, with a great deal of show, brouhaha, television interviews, and novels, and, of course, these newly minted Pascals are a dime a dozen and will soon be forgotten.[5] All of these are assertions of freedom. The artist is ready to do anything except to produce a work of art. He will be a thinker or a technician, a solver of problems, whose freedom consists in challenging everything and posing dilemmas. We have already seen this attitude with Dubuffet. Rauschenberg, one of the creators of Pop art also began by "challenging everything" (his first canvases were simple white spaces without design or color), and then he began to produce works that challenged consumer society, and from there on, he encountered "problems," "physical problems that arise from the manipulation of objects that mutually affect one another."

But mathematical musicians also present their productions under the guise of problems, seeking to create or to find the adequate tool. Barbaud will use a specific form of mathematics suitable for composing music, but he will base his work on mathematics, which he respects through what it offers. Xenakis will use mathematics, which he considers as ancillary but he will modify, according to his own inspiration, the results that mathematics offer. Jeantel will study the way in which music is composed by using mathematics as a tool of understanding. The most important issue rests on the problem that one poses and on the manner in which it is resolved. The act of composing challenges the existence of the work itself. The emphasis on compositional *technique* effaces the work. The technical mastery of the artist leaves only silence. But another expression of the artist's sovereign freedom resides in his claim to be a savior. Not only committed to a cause, he considers himself a savior and a magician. The artist believes in his sovereign right to change the world—not only symbolically but also realistically. All that is needed is poetic expression, the production of a poem, for the world to be changed, for the revolution to be accomplished. G. Mathieu announces bombastically that "salvation is and can only ever be in the artist since his natural vocation is to create forms from which will spring spontaneously a language that expresses the profound and latent aspirations of the community …" The artist creates mutual understanding among men and a harmony between man and his environment. Nothing less: he is the true demiurge. But, even more remarkable, he is the "solver" of problems, the creator and savior, the inventor of "language forms," who is completely dominated by the total subjectivity of the artist, and by this fact, the artist appears as sovereign and free with regard to our society and our culture. He creates his own language. He produces his own individual forms (whether they are recognizable or not is unimportant). Above all he insists on expressing himself. Even the architect wants, first and foremost, to express himself. Malraux was correct to state that "modern art begins at the moment when Manet decides to be everything and when his model is reduced to nothing."[6] The artist commandeers creative space; what matters is what he miraculously makes. Since this absolute pronouncement means nothing, the artist in his sovereignty can claim his freedom to be beyond understanding. Whatever position he takes (art with a message or formal art), this freedom that no one else possesses empowers the uncontested critic of the entire social system. In whatever this artist says, makes, produces, or does not produce, he "challenges" all of society. All have been inured by this endless questioning and by this dedication to subversion. The artist sees himself an exile who is separated by an unbridgeable gulf from the current noxious society. Every artist is a poet or painter condemned. The artist's lot, full of blood, sweat, and tears, in the eighteenth and nineteenth centuries, has become a meek chest of drawers. From the moment that one becomes an artist, one is *ipso facto*, a condemned artist. It goes without saying that this poisonous Capitalist society (or Soviet Socialist, which comes down to the same thing) can only destroy the free and spontaneous creator. The Chattertons, the Malfilâtres, abound, but they don't starve in an attic; everyone knows them thanks to television. It's a fight to the death between free artists and a society made up of television viewers. But this free and condemned artist enjoys all the advantages of society.

Edward Albee creates, in his *Quotations from Chairman Mao Tse-Tung*, a supposedly revolutionary play in the middle of the Vietnam war; he attacks American imperialism and proves that Mao is the only bearer of truth. And this revolutionary piece was performed at the Buffalo Festival in New York, without problem, censure, or prosecution, and all who went congratulated themselves on their prodigious acts of non-conformity and freedom. The outsider who challenges is front and center when it comes to aesthetic, theatrical, popular, and financial success. Just the same, consider this: either one *truly* attacks society—and yet is awarded for doing so—or else the attack is carried out in a void of appearance and pretense, for which society will pay to avoid facing the hard questions. But this is of no concern for our artists: they have it all, a good revolutionary conscience and social success. This did not happen yesterday. Benda had already shown in *The Betrayal of the Intellectuals* that intellectuals and artists were selling themselves to the system. But the new wrinkle is that, by incarnating the free man, the artist becomes a bureaucrat and that by being a licensed opponent he collects all the advantages. He contradicts a society chained by an art that is unchained, but this unchaining is a simple psychological release, a very useful safety valve that retails the ideas that are already common currency. One can recall Piaget's cogent ideas on the subject of anticipation when he explains that anticipation is part of the biological order. Embryonic development is nothing but the systematic anticipation of later states and functions. And so this is right up the artist's alley; he has that function—in no way revolutionary—of anticipating what is already present invariably in the society to which it will give birth. And so he dramatizes and tragedifies, seasons to taste, and adds an ejaculatory shudder to a production that is little more than mechanical in a society that is more and more mechanized by the megamachine. It's the successful man who can speak with authority about subversion—Pierre Boulez, for example. Honored by all authorities, decorated with all awards, welcomed into all circles, he complacently announces that one must spread subversion everywhere. One can only wonder if this is the cliché to beat all clichés. That is, subversion, revolution, and so on, are the conformities of our time just as order, morality, and beauty were in 1860. The alliance between art that calls for absolute artistic freedom and the monied establishment has become established law. One need only remember the hilarious mini-scandal that erupted when a minister of culture declared that he would refuse to back artists who spat upon the government and shouted Revolution. Let's be clear: either you get money or the right to challenge, but not both. You can imagine the hue and cry that was provoked. "But we absolutely want both." Absolute success and at the same time a good revolutionary conscience. Anti-art and anti-literature, like anti-society, draw their abundant technological and financial resources from those they profess to deny and to defy. We want to be paid by the society that we challenge. We are bureaucrats who make a show of freedom and are the chorus of free men in *Ubu in Chains*.

Like Boulez, Michel Butor is a specialist in the universal success of challenging. America covers him with flowers and the Ford Foundation welcomes him in Berlin. He is not only an authority in the world of arts and letters but also in that of society in general. Now, it goes without saying that this Anointed One is a challenger of the absolute. I do not mean to imply that he is a social reformer but only that he, and those like him, totally agree with the social structures and general directions of that society. Pop artists, those so-called revolutionaries, are not simply involved in the Capitalist world by earning money: they are already, by their very activity and their concept of art, cheerleaders of consumer society masquerading as critics, just as pop musicians are the destroyers of any possible resistance to the social system. The "image icons," far from raising consciousness, have produced a supplementary adaptation to the religion of consumerism. Not only has art become an economic production, a piece of merchandise, but art that challenges and proclaims freedom fulfills an essential function: it's the label, the guarantee, that this wonderful society is indeed free because it sanctifies the heroes of freedom and revolution for the admiration of the masses. It is commendable that the modern artist lends himself to this game so easily and that he assumes this role and destiny. And this strange entity, namely society, endowed with an uncommon sensibility, with an infallible instinct, knows what's up. It cannot live without heroes, saints, martyrs, individuals, and exemplary free men. On the condition that all of this is "for show," for window dressing, for a parade to amuse the gawkers, the artist, through his research, his technique, his use of materials and devices, his mathematizations, his elimination of meaning, the artist finds himself in fundamental agreement with the basic premises and framework of that society. On these grounds, the agreement is accomplished miraculously. The artist plays on two stages. Thus, the challenger is decorated with the Legion of Honor and receives the Nobel Prize. He facilitates the future. He joins the avant-garde. That is to say, he proclaims the conformity *du jour*.

The ideology of the avant-garde (and certainly neither Rimbaud, Lautréamont, nor Van Gogh, thought themselves a part of it) implies precisely this point of view: keep one step ahead, no more, on the road that most of the flock is following. But the avant-garde does not invent or lay out the road; it follows slavishly orders from headquarters! One could say that the true avant-garde is the one that does not know or recognize itself until after the fact; interestingly, now, all the "true" modern artists claim to be avant-garde and want to break with the past and with society in general. Each one wants to be the political, ethical, and aesthetic protestor who claims to live out a unique adventure but who, in reality, expresses nothing more than the technical nature of his time. For this reason, despite appearances, there are no avant-garde riots. There is no longer anything like the premier of *Hernani* and *Tannhäuser*.

Jacquart defines the avant-garde in the theater by its will for total opposition and rupture with regard to the past, and, then, by its revolt against good taste, all of which is expressed by its aggressive attitude for the public. But he is correct to speak of "seventy years of avant-garde." It's basically the problem of youth: the most revolutionary young people grow old. The avant-garde of 1900 is the rear guard of today, consisting of the most conservative members of the flock! But who today dares to oppose the future, to oppose progress? The avant-garde is the creator of what has never before been seen or heard (which must therefore be right), but which can and will of necessity be produced and reproduced by the computer—the supreme expression of the artist's freedom.[7]

In the presence of these endlessly repeated claims that boast success, we must proceed to an analysis that is a little more rigorous. The modern artist, far from being free, individualistic, avant-garde, and protesting, belongs strictly to a clearly defined milieu, perfectly determined. Ultimately the being of the artist is denied, as we have seen, by the very theory of modern art that is a result of the technical system, which he expresses. The avant-garde constitutes a sociologically conditioned milieu with no surprises. The avant-garde (which, we recall, includes without exception all the renowned artists of today) is centered within the aesthetic milieu and the intelligentsia with its own vocabulary and passwords, its points of intersection and relationships, its salons and patrons (Madame F. Gould, for example, in 1946). This situation it not new, clearly, and is not unique to artists in our own society, although the modern artist is more connected to this milieu than in the past. When one is of the "milieu," one's career is assured. Out of one hundred examples I'll offer one: an unknown author will never be published in high-class editions. An author who belongs to the milieu, who is a member of the fraternity, will enjoy incredible benefits: at considerable expense, he will get texts in different colors with brand new signs and symbols. And, what is this? Because huge printings are assured? Not at all. The author is a member of a fraternity and, therefore, his aesthetic whims must be followed. Furthermore, under his aegis, all great works are composed or painted. If so and so can write that, then I can write this. The imaginary museum counts for much less than the interrelationship between members of this milieu, which has less to do with Rembrandt and Bach and more to do with the little pal I met at a cocktail party. La Casse and Le Senne offer another interesting example. A novelist like La Casse gets enthusiastic reviews from one of the members of the milieu in what amounts to a case of "I'll scratch your back if you scratch mine." This practice is so evident that there's no use hammering away at it. Let us simply note that this seemingly independent and sovereign author exists only as a function of a very precise, determining, and narrow milieu, which imposes on him a code from which he draws the totality of his inspiration. The wildest painter, the most contrary poet, the most abstract musician frequents all the best Parisian salons where imprimaturs will be conferred, received, and exploited. The judgment of the milieu is wonderfully efficient, and when one considers the schools, the styles, the latest creations, etc., one must first ask: who has been schmoozing around with whom and who has been invited where?

Let's leave all this aside for it has nothing to do with technical society and it only shows that the conditioning of our modern artists is extensive and that they undergo, simultaneously, both a cumulative conditioning by the technical system and a *traditional* conditioning in the salons, the fraternities, and the "milieu" in a word. In other words, this forest god fallen from heaven into a boundless freedom is completely shaped and formed by these two influences. To this one must add the conditioning of the public and patrons. The development of freedom in the modern artist is often described as a transition from the patron to the public. But the public (which is fashioned by the critic and by the technical system) is deeply influenced by publishers, concert organizers, and gallery owners … And this constitutes the totalitarian, exclusive, and deeply rooted characteristic of this conditioning. I'm astounded when they announce that nowadays there are no longer artists of unknown genius. All depends on commercial decisions in the capitalist system. But, in the socialist system it's worse; all depends on ideological conformity controlled by a single source. Integrated into this society, art is necessarily split between capitalism (this being the major player) and socialism. Capitalism is the artistic stock exchange. In addition to the long-standing problem of the museum as a canning jar of traditional works we must also note another relation with modern society: the museum as a result of economic expansion—the work of art as merchandise and speculation. (J. Michel gives some mind-boggling examples in his remarkable investigation: "Le jeu de l'art et du hasard," *Le Monde,* March,

1976.) The demand for production, especially with the creation of modern museums of art, occurs in all nations undergoing economic expansion: every museum of modern art that opens creates a potential market for several thousand works of art. Although it happens occasionally that, as J. Michel points out, there is more money than works of art and that the advertising business creates the artist: "Once inserted into the process of the history of art by exhibitions, books, articles, broadcasts (of interviews, of opening shows), in short, the entire kit and caboodle of cultural communications ... the price of the works takes off." And it is that process that consecrates the work as a work of art and the artist along with it! The artist as a mythical personage is a producer of market value in capitalist countries to be sure. But why is there practically nothing produced in socialist countries? Problem solved! In socialist countries the consumer base hardly exists. Now, what is true of painting is also true of music: every "economic miracle" goes hand in hand with a "musical miracle," in the United States, in Germany, in the Netherlands, in Spain, in Japan. Modern art is tied to capitalism. Everybody knows this. But, more deeply than anyone realizes. And what interests us here, and we will deal with it later, is its relationship to *technique*. J. Michel puts it well: " A society haunted by the robotic nature of mechanistic civilization in the midst of upheavals in cultural production, gazes at itself in a mirror of truth with cleansing powers." Either the State or big industry are the two current patrons presently. The new headquarters of the Renault factories at Billancourt is a good example: it is a veritable museum of contemporary art. The description given by J. Michel is very remarkable:

> *The great entry hall is a monumental kinetic sculpture, which plunges the visitor into the heart of a total artistic environment. On the ceiling a white rain of thin plexiglas rods; on the walls, grooves and squares, in black, blue, and white, a kind of sculpted writing that constantly vibrates. ... We do not gaze at this work of art: it is the work of art that enfolds us from all directions. We are inside.*
>
> *On other floors one encounters two frescoes by Arman, the artist of industrial society, whose "Accumulations" transform water pumps and cylinder heads cut into strips into decorative motifs. In the computer rooms we see impressive mural paintings on enameled sheet metal by Jean Dewasne. The employees' cafeterias are decorated with kinetic works by Soto, Leparc ... and, to finish things off, a surprising suite of executive dining rooms at the pinnacle of the tower.*
>
> *This is the floor of luxury: chrome, black leather, white marble in the tradition of the interior architecture defined by Mies van der Rohe. The series of decorative tableaux by Vasarely on polychrome metal seem to revive an atmosphere of Berlin Bauhaus at the moment when design and industrial art were born.*

In other words, we have a purely decorative art, adapted to the industrial world, which expresses (correctly) the technical milieu and by this fact introduces no rupture, no conflict, no dialectic, no counterpoint: it is truly the perfect illustration of my thesis of an art expressing *technique*—integrated into *technique* and designed to integrate man into this process of enculturation. And this is truly the artist's task. There is an interesting and moving interview with the American painter Robert Motherwell (*Le Monde*, June, 1977) where he idealizes this process no end and speaks about snatching out a visual order from the chaos that surrounds us to become again like a little child in creative spontaneity, to be a whole man, etc. But then, and he mentions the conditions in which he works, he recognizes with complete honesty that he is a type of businessman, with a team of four assistants, a full time secretary, nine studios which turn out his paintings, and hundreds of telephone calls per day. Angelism is transformed into a techno-commercial enterprise.

It is certainly not the "public" that determines good taste. It is the dealer or *apparatchik* who evaluates what is good for the public and for which public. This method is not foolproof, and there are as many chances that an author of genius will remain completely unknown. But the conditioning by the patron class joins with that of the "public." Nearly all that translates into the technical impulse is hard to market. Avant-garde sculpture or painting marginalizes the "public" that will not buy it. Commissioned works fill the void. Most often, it is institutions that finance the commissions. The one percent who are the French aesthetic establishment rely on painters and sculptors who are paid by the State or by huge businesses.

In the nineteenth century the Fauves, the artists who showed supreme contempt for the bourgeoisie (but who lived off them) were completely co-opted, we now know, by the capitalist bourgeois system. It is the same today with the technical system. The common grandeur of the relation between artist and patron is in no way aesthetic but is exclusively technological. One must also make way for the "producer." As Schaeffer clearly points out, "The artistic creator has lost, if not his independence, at least his initiative, and the latter belongs to the producer who is inspired by a demand for the product," and this comes from statistics and commercial evaluations. The technical milieu does not only impose the means for the work (and one must recall that the more the means are complex and costly, the less there is artistic freedom). Even greater restrictions are placed on art by the prevailing preconceptions of the technical milieu. Now, these preconceptions derived from the technical process are shared by artists and patrons. Those who do not honor these preconceptions cease to be interesting and to have an aesthetic future. *Technique* imposes its laws on architecture both in the constructive and economic sense. But the same goes for all the other arts, which absorb the prevailing notions of *technique* in their approach to conceptualizing things and actions.

Today, there is no longer a sovereign demiurge who emerges freely and manipulates the means in its surrounding universe. The artist is as narrowly preconditioned as the cosmonaut in his capsule. He knows exactly what he must do in each case and in each moment, because the technical imperative has so deeply penetrated his soul that he can express nothing else. Mentally determined to this degree and driven by the ideological imperative of commercial success, the artist has no room to express his freedom: he produces what best fits the commercial and technical demands, and this is completely in line with what we were saying about subjectivity. Subjectivity does not exclude the demands of *technique*. If the painter does not directly express the technical system, he falls back (because it is the only path left) into a pure "internal" subjectivity, no less conditioned by the system. Abstract art reveals the internal landscape of the painter, who has always expressed a relation to a milieu (even when it did not involve standard figurative art). But, if the painter refuses to represent directly this new technical milieu, he is left to himself to display the interior adventure of his personality in its "pure state," with no clear impression of the world in which he lives. In other words, one has a picture of the psyche. Hence, these painters reveal a being that is totally self absorbed and isolated. Thus, *technique*, at the same time, returns to a subjectivity. And here we find that the role assigned to the artist is that of a man with a personality in a depersonalized society resulting in the suppression of the artist as subject. Not only is this free man perfectly conditioned at all levels more than ever because of the growth of *technique*, but he also has less self-awareness than other men. However, this co-opted, nonexistent author, perfectly established and conditioned by the technical milieu, like Calvino's non-existent knight, with perfect efficiency, nonetheless, collects his author's wages and becomes extremely angry if his kingly creativity is not honored (which, of course, is not creativity but only an anonymous game!).

Thus, on the one hand, we have the disappearance of the artist behind his instruments and his collectivity, which exactly mirrors man in the technical world. How could art still be a message from one individual to another when machines and their works no longer have anything to do with the individual? But, on the other hand, these propositions that perfectly conform to the structure of technological society are refuted in practice by their promoters who do anything but disappear into anonymity. They continue to demonstrate the same typical acts of pettiness like filmstars and personalities who are thin skinned, jealous, egotistical, social climbing, complaining, and depressive; oh! how human they still are. But they are incapable of giving freedom any other image than this monkey business. *Technique* imprisons them more and more in their recognized and patented role of artist, which implies a type of temperament that celebrates the elimination of the kind of man they pretend to be—joyous prophets spreading the news like latter-day troubadours.

II / *The Necessity of the Art Critic*

In this vast universe, the art critic finally achieves his principal role. The artist is only a secondary element in relation to the critic who makes and unmakes styles and reputations. We will not address the general question of the critic but only his situation in art as it relates to technological society. However, we will not attempt to develop the no-holds-barred account of Daix who, basically, demonstrates that the best, the most intelligent, the most progressive, the most clear-sighted critics *are always wrong*: and he demonstrates this for Baudelaire and Zola in their understanding of art! The argument is interesting and is food for thought when one considers that it is the critic who *makes* modern art, but this is not directly linked to our theme. Let us allow the reader to consider this topic. Let us note that the art critic is a recent development. This man who is a scholar of the material in question, music or novel, painting or poetry, who knows all that can be known, who is the true expert in all his knowledge of the imaginary museum; that man, a specialist, a meticulous connoisseur of all the techniques, is incapable of producing anything by himself, but, as the occupant of the public podium, he makes his opinions known. He promulgates evaluations, and he reveals the philosophy and meaning of these works. He decrees what is good for the general welfare, the culture, and what will be the legacy of our present world for the future. He can add nothing to this legacy but his *explanations.* The art critic did not exist in the seventeenth century, although there were a few hints.[8] In reality, he is a product of bourgeois, industrial, mass society; he is a shareholder in the culture, whose conscious and willful reality originates in the same era (along with the idea of culture), as Charbonneau has demonstrated. The critic owes his existence to the mutation of the bourgeoisie: the bourgeois, perhaps uncultured, harried and involved with other needs, and dedicated to utility, does not possess the same understanding of art as the aristocrat. For the latter, there was no need for explanation. By contrast, the bourgeois, the philistine of the Gilded Age, needed explanations, needed to be led to understanding. And, just as businessmen needed their brokers, so, in matters of art, the bourgeois needed their critics in order to discern what kind of art to buy. And for the Bourgeois buying art is an act of status. He must not make a mistake. First and foremost, the critic guarantees the durability and lasting value of the work in question. The critic is just another business agent whose job is to guarantee status.[9] Finally, the latest role of the art critic is linked to the development of mass society. Art is no longer the purview of the privileged, who, in the salon of a wealthy patron, will hear Mozart play his latest symphony. No, it is the broad public audience, which appeared in the nineteenth century and is oriented toward large public gatherings. One must inform them in advance, one must warn them of what they risk hearing or seeing in a theater or exhibition; one must encourage or discourage their attendance; one must then allow them to gather together and to know on the next day what they have seen and heard. The art critic is a publicity agent for modern art.

When a mass culture, directed from above, develops, it is necessary to have an intermediary who utters an opinion for an uncertain and undecided public in the face of the new. Hence, mass culture must continuously offer new products. The role of the mediator is, therefore inevitably rigorous and fundamental. He plays the role so well described by J. Monod, in *Le hasard et la nécessité*. There are a thousand artistic experiments thrown out at random, among which the critic sanctifies one or two that, by this action, obtain durability and found a new style. I know that in their great modesty critics clamor that they do not have so much power! But this is the way things are. All "rebel artists" have had a certain number of critics as their supporters. This is increasingly true as the critic, well-schooled by the errors of his predecessors who failed to recognize the art of tomorrow in the work of today, is carried along by the ideology of progress, which clothes him from head to toe, binding him more and more to the benefits of modern Technique. He bets on the future, that is, he rejects in advance anything that recalls the past in order to strike out into the swamp of the new and the most absolute, which alone appears worthy of interest. Be that as it may, the critic is, therefore, the specialist in discourse on Art. In this discourse the critic determines and defines style. But this is an aspect of the society of the identical (Bellini), where the identical differentiates itself from other moments or from other things. In the nineteenth century the critic defined "good taste," which is now despised, but his role is still the same, and, currently, he defines the concept of "right thinking" that challenges conformity and pushes the limits of toleration and creates what appeals to the public. The critic, as always, defines a style drawn from an ocean of apparently diverse forms and announces what is right and what is to be followed. Thus the critic makes and unmakes art by defining to a potential clientele the boundaries of originality. For the general public spontaneity is not possible in the technological society. The link between the discourse of the critic and art itself is so essential that it appears, for example, in the view of Abraham Moles, as a proof of art's vitality. Everywhere they have proclaimed the death of Art, he asserts; now we are witnessing an "unprecedented flourishing of *doctrines* and *movements*," which prove that art is alive. However, these doctrines are the work of critics. They produce an infinite amount of discourse on art, but one must remember that this is not art. And, we must ask: is grandiloquent discourse about an object a sign of the object's absence? Freedom and justice are exalted in the political sphere only when they are missing. The wonderful flourishing of hundreds of types of criticism, each one more refined, intelligent, and erudite than the others, each rivaling the other for audacity and profundity, is the veil behind which pure and simple absence hides. But, finally, we have not undertaken in this analysis to discover if modern art is really art, and at this point, we will not run down this dead-end street.

The critic is the specialist in the discourse on Art. But, there are in reality three different types of discourse according to the tendencies of art and also in the dispensations where art develops. From this fact, there are three diverging roles for the art critic. In one case, the critic makes clear the message and in another what the message is not, and this occurs in dispensations dedicated to preserving orthodoxy. We have seen that in an early stage, the modern artist wants to convey a message, but it is no longer possible simply to say what he has to say. This would not be art. The abrupt and simplistic approach of Aubigné's *Les Tragiques* is no longer our game. The message must be so complex and express the infinite complexity of our society and be presented in such a subtle way that tickles our interest, perhaps, but is not clear-cut. Furthermore, the message has become so subtle that a translator is required to explain to the public whether there is a message or not. The profundity of Fermigier is needed to make us understand the dramatic sentiment, the suffering, the exile, the weight of the negative, for example, in the work of Tàpies.

Clearly minimalist and post-minimalist paintings and sculptures are nothing, absolutely nothing, without explanatory discourse. We are told that it is a "mental" art, which now requires conceptualization and no longer the sentimentality that has ruled art for too long. I can buy that. But I do not see in what way a red X traced on a white sheet is an any way "conceptual." Now, I must explain. A work like this has no character or any intellectually discernible quality unless the artist or the master know-it-all steps up and reveals the intellectual process, the means of understanding, and the logic of the work. This is what we could call an "instruction manual of poetics." I'll buy that too. But, why should this act of drawing two bars on paper be a greater act of creation than that of a lathe operator in his workshop? Here we are in the domain of intention, of magical transfiguration, revealed through discourse (even though one condemns this logical use of words, one has no other recourse to explain the abstruse schematics presented by a modern painter!). As for saying that this art expresses a primitive sense of being or a need for aesthetic fiction ... no one, unless he ruminates on the critic's philosophical explanation, is affected by this so-called aesthetic fiction, which unfortunately is only an actual reflection of what is most dehumanized and most technological in our society.

It would be an endless task to attempt to recall every case in which the critic has revealed the artist's message to the public.[10] This can be said of the cinema as well as of the theater. For we no longer tolerate theater with a thesis like that of Camus and Sartre with their well-laid messages (and, nevertheless, already quite convoluted and recondite, in relation to theater with a message, like that of Ibsen in 1900), we accept that there is a message but only one as obscure as life itself. And so, only the talented critic can translate the untranslatable Beckett, etc. There is no mystery here, and we won't insist on it. We know the message is there. Sometimes a sign informs us that there is a message, but we have no sign of what it is. A scary jumping jack with a cigar in its mouth appears on the stage of Bread and Puppet. The know-it-all specialist lays it out for us in explicit form, and we are satisfied to have participated in the double play of concealment and revelation, Art and the explication of Art, which also is a demonstration both of an intellectual and revolutionary spirit. All's well that ends well is the critic's guarantee when he comes to unravel the message for us.

The critic's work becomes much more subtle when it approaches informal art that refuses to say anything.[11] Here we would be left fully silent and alone if it were not for the critic. "The fascination with the difficulty, indeed the incomprehensibility of works of art, betrays the desire to discover a heretofore new secret, an unknown meaning in the world and in human existence. One dreams of being *initiated*, of penetrating the occult meaning of all these destructions of artistic language, of all *originating* experiences, which seem, at first view, to have nothing in common with art." (Mircea Eliade). Thus, we are in the presence of an esoteric, cryptographic art, which refuses and challenges all meaning and which refuses to recognize that there is anything to say. And the unbelievably ambiguous role of the critic consists of finding meaning in the object placed before him, of revealing the sense in its nonsense, of charting the symbols in this absurd playroom cacophony, while all the time denying that spoken discourse has the least rationality and, willy-nilly, extracting something more or less comprehensible from these completely enigmatic and intentionally nonsensical works. If the critic were not there, the public would be as completely absent as the meaning itself.[12] Basically, the critic plays the role of a tour guide for a public on quick visits to museums or the role of a barker for the jumbled images that, in 1900, were shown at magic lantern fairs, or the slightly more noble role of the priest explaining in the Middle Ages the meaning of the *libri idiotarum*! Each time there is a gap between what is to be seen, understood, appreciated, or retained, and the aptitude or capability, the knowledge, the culture, the sensibility of the one who is to appreciate and retain … the critic fills in the gap. The work no longer speaks for itself; the critic speaks in its place and situates the work in the great current that carries art to this point. He becomes the irreplaceable companion on whom the artist relies. The same goes for subjectivism. It is well and good that the artist plays at expressing only his unique experience, his exclusive and irreplaceable subjectivity, but if he did only that, he would not be understood. Fortunately, the critic is there and will explicate this subjectivity with the help of appropriate techniques and will explain the artist in such a way that the general public will be able to participate in the artist's unique nature. Thus the critic decrypts the work of the text hidden by the author. He erases (theoretically at least!) the author in order to leave the text in its pure state, or else he unearths the unconscious message of the work; he interprets "what is said but not said," and reveals the meaning of the unsaid by saying it. We have gone far beyond psychoanalysis.

I am not at all sure that modern art would have evolved as it has if it were not for the critic. In other words, it is an enormous resource to know that an explicator is behind the artist and whatever he produces; the explicator will take charge of providing a meaning and of decrypting symbols even when none exists. This will happen provided the artist in question falls in line with society and with the current development of sensibilities in artistic milieus. If this happens, the artist no longer has to worry about expressing a "form" attached to a "meaning," and he has the rare privilege of declaring that there is no meaning, that art is pure form; he knows that the critic is the specialist in discovering sense and nonsense, a meaning in pure form that pretends to say nothing. The critic is nothing more than a safety net for a high wire act. The artist, according to the model granted him by modern society, has the added privilege (and one that he always claims) of insulting without consequence the philistine public off which he lives. The critic is there to exalt these insults and to demonstrate that they are really a form of respect for the spectator who is involved in a newly minted reality. But, in order to play this role, the critic becomes a technician of art. He knows its history, its techniques, and its resources better than any artist. And the more the work becomes abstract, the more a technician is required who not only displays his knowledge of *technique* but, in addition, can elaborate his own techniques for interpretation. He possesses instruments the others do not. The critic, after having become an indispensible character of capitalist bourgeois society, becomes a representative epigone of the technical system. He makes art and the comprehension of art a *technique*; he leads us to think that there is no such thing as untutored reading or vision (except in those cases where it is false); he establishes himself as the sole judge of success and value, which is precisely the role of the technician. The artist brings the critic a piece of raw material, and the critic, by means of his technique, transforms this material to make it useful, and he sets up the machine for providing maximum efficiency in its impact on the public. The critic, in his current role, and especially in relation to abstract art, fills a social function of monopolistic mediation based on technical competence. He is the technician for arts and letters, comparable to MBAs and other technocrats. And, like them all, he reduces every art form to an ensemble of techniques. It is not by accident that the denial of meaning and a refusal of anything to say is the result of the ping-pong match between critics and artists. The critic assures his pre-eminence, his social role, his indispensable status by asserting that there is nothing in art which is not hermetic, nothing that is not symbolic in the second or third degree, but that everything is beyond the understanding of the lay person. And the artist plays the same game, assured that his technique will hoodwink the public.[13] From all evidence this technocratic role excludes meaning, because meaning escapes *technique* and can be understood without *technique*. Informal and abstract art fit completely with the aims of the critic, which is why the famous quarrel about the *nouvelle critique* seems entirely irrelevant to me. It could not be otherwise: in a technological society the critic, of necessity, becomes a technician who reduces art to form and cryptography.[14] The debate centers on the very validity of the technical system. The *nouvelle critique* is nothing but a literary and cultural translation of social technocracy, which brings nothing new and, above all, nothing revolutionary, and Roland Barthes's claim to create a science of literature is nothing more than an application of a form of technology. In this debate, moreover, we get too bogged down in the French *nouvelle critique*; twenty years earlier, if I'm not mistaken, the movement was launched in the United States. John Crowe Ransom, in *The New Criticism* held that the work must be considered by itself, "independent of all historic, social, ideological or biographical consideration … The novelist's subject is neither society, nor psychology, nor ideology, but rather the word and the structures of the sentence. He seeks to discern in the formal structures of the work its essential traits … The symbol is no longer a codified image to be deciphered. On the contrary, one must ferret out the plurality of possible ambiguities or multifaceted language as Barthes would say later on." (J. Cabaut). French debate on New Criticism has at least had the advantage of illuminating the fact that there are many tendencies in this literary current. For, on the one hand, we have the claim that there is no meaning and that the word speaks for itself in an autonomous language that exists beyond any creator, but, on the other hand, in Doubrovsky we have the desire to be free from a single language and to find linkage between writing and lived experience, to explain the work as it applies to every man, and, what is more surprising, we have Pingaud who claims that the aim of criticism is ultimately true meaning. In all cases what the author says is never to be confused with what he wants or would want to say. And it is New Criticism that seizes on an author's real meaning behind what he says. But despite this difference, we conclude with two quite remarkable positions. The first position is the pre-eminent role of the critic, who is situated above the creative artist. Roland Barthes states this position cogently; there is a kind of hierarchy where one goes from the word to the language, from the language to the work, and from there to the criticism, which is the "ultimate act of literary creation, a symbol superimposed on the symbol created by the work." Without the critic the work is sterile, because the critic is essentially the model for the reader; and so, the work must be studied not as an extension of the one who creates it but of the one who reads it. Reading constitutes the work. Hence, criticism is the fine flower, the ultimate point, the supreme creation of the artistic process. This is quite astonishing, because this concept must be applied across the board. But the second position is no less interesting: New Criticism consists of a body of doctrines and methods so overwhelming that authors under its sway dare not challenge it under pain of being thrown into that swamp to which Barthes consigns the average reader for whom he has only contempt (and from a good leftist, what else do you expect?). In reality the new critic imposes his view of things on the author, and the author begins to write as the new critic demands simply to prove the value of New Criticism. This interesting development tends to confirm what we have suggested before, that the artist operates under certain external technical constraints. Be that as it may, the critic (technocrat) is now the driver of style and creation.

Finally, the critic plays another essential role but this time in countries of Socialist ideologies (I've already mentioned this role in what I have said recently.) He is the guardian of ideological correctness in the work. He is the guarantor of orthodoxy. And so his only role is to certify a given work's ideological correctness to be handed over to the public. Hence, the work's fate depends exclusively on the critic's verdict. But in order to successfully carry out their function, critics determine the general line appropriate to follow. They become the mediators between power and the artist, while claiming to be mediators of the social body. This depends not only on the ideological climate but also on the fact that any work of art is destined to fulfill a certain social role with its pedagogical and propagandistic value and is important to the extent it serves to build socialism. The work of art is integrated into the social body, less through its beginnings than through its goal, and the critic is the guarantor of this goal and integration. It is therefore clear through these diverse functions that the critic has become the most important person in the world of art, and this relates at all levels to the technicization of society.

Thus, it is amusing to note that New Criticism violently attacks old criticism proclaiming that the old critic was an agent of social control, the ultimate police force created by society to monitor expression of thought and conversation about values … but those who exercise this function in a terroristic manner are none other than the new critics, much more so than the traditional university critics! But, in their exclusive role understood by the new critics and as members of a befuddled aristocracy, the critic has become society's policeman. But, if traditional critics were the police of traditional society, then the new critics are the enforcers for technological society with a deep-rooted and exclusive attention to everything that might threaten this society. The more the current critic appears to be avant-garde, the more he is in fact the preserver of technological power and its laws of aristocratic aesthetic creation. And, finally, the artists who are convinced that they will never say what they claim to say, that they will never express what they wish to express, or, that there is nothing to say or to express, will, nonetheless, write and paint whatever and then simply hand it over to the technological guru. We are witnessing a very interesting fact: the critic makes up for the lack of meaning in the work, while simultaneously offering a technological paradigm of precise rules and a metaphysics of art and language *all of which is nothing more than the development of the technological system in relation to the natural world.*

Notes

1 This paragraph on the artist as specialist in freedom and as a person owes much to the remarkable book of Bernard Charbonneau, *Le paradoxe de la culture*, 1960.

2 Of course freedom can be expressed in virtually any arena: for example, in the case of André Steiger's theater group The Tact, the director examines the play to direct and then begins to make cuts, to rearrange, or even to limit what will be performed in a scene. With considerable attention to the *mise en scène* ... This tries to escape the terrorism of the author and the written text but, as often happens, one substitutes the terrorism of the actors and of the *mise en scène*!

3 Francastel, *op. cit.*, p. 216 and following.

4 It is necessary to emphasize that we are reaching, at this point, another way to describe the contradiction and rendering of modern art. We have studied at length in the preceding chapter all that constitutes the disappearance of the subject in art, but here we find a contrary and complementary current in the exaltation of the artist's personality. No longer is anyone an artist who creates in his own right, but the artist is the model of the absolutely free person. Once again this is an expression of the juxtaposition of contraries without reconciliation or synthesis, which is the dominant characteristic of the technical system.

5 *cf.* J. Ellul, "De l'inconséquence," in *Mélanges en l'honneur de D. de Rougemont*, 1977.

6 And G. Picon's denial of this point is worth absolutely nothing.

7 Consider an interesting poll from the magazine *Digraphe* (1975) on the avant-garde as seen by two hundred writers. What stands out is at first a collection of declarations on the characteristics of the avant-garde today and a general agreement to consider it as political. It is true that the criteria of this political avant-garde point to a totally worn out rear-garde of the left, because the avant-garde still remains close to the worst of the rear-garde, only allowing a small play for adaptation to a new situation without questioning or challenging anything. But by its political nature, the avant-garde is in full accord with mass art. Lebot recalls (while criticizing it) the stance according to which the avant-garde corresponds to mass art, to the ideological indoctrination of the masses, and is the ideological agent of mass media. To my mind this is precisely the role of the avant-garde.

A second set of declarations sees in the avant-garde a mysterious unknown society capable of challenging everything but which ceases to be avant-garde once it is known: this common banality collides with the imaginary museum and with the mass media. If the artists and writers in question are truly hidden, they play no exact role in relation to the work of mass acculturation. Hence, the industry of the spectacle and the ideology of progress generate an endless search for avant-garde talent.

Among the texts on the avant-garde only that of Michel Tournier offers something interesting; here are two extracts:

> Let's open the Littré *dictionary*: "Avant-garde. Marine term. Old building placed at the entrance to a harbor for surveillance." That says it all: a rickety old boat leaking everywhere, playing the role of a cop at the entrance of an old folk's home ...
>
> If by accident the avant-garde were to assume a political function, it would create the most reactionary reign of terror in the name of revolution. We've seen it and we'll see it again.
>
> On the other hand, in a totally different sense, and treating the avant-garde seriously, Estivals (et al.), L'avant-garde culturelle parisienne depuis 1945 *(1962).*

8 Indeed, I am aware that the first activity of art criticism is Italy in the fifteenth century, but this did not play at all the same role as today.

9 Of course, I refer here to art criticism in the narrow sense and not as Bellini does, for example, when he sees criticism as a permanent social function, which consists in measuring art in relation to the values of an age, but when he correctly notes that art criticism is, in reality, the basis for a history of art, which, in turn, is nothing more than a history of art criticism (since necessarily it's a question of the relationship between art and value). He returns, unconsciously, to the modern age, since the concern for establishing a history is, ultimately, a preoccupation of our society.

10 One can cite as an example of monumental exegesis the catalogue of Marcel Duchamp's incoherent work in four thick volumes.

11 We have seen previously that, whether in the case of art with a message (p. 186) or in the case of theoretical art (p. 221), the artist explains himself obliquely through his work: and here we see him largely aided and supported in these two endeavors by the critic.

12 The prodigious amount of useless brainpower that these critics embody never ceases to amaze me. Among a thousand possible examples, one can cite Forge's disquisition on the subject of Rauschenberg's *Allegory* (exhibition of modern art, Amsterdam, 1968) in which he combines an umbrella and pieces of metal: "This is one of Rauschenberg's most characteristic works in his desire to shock. What is extraordinary is the way in which forms as powerful and aggressive as metal and an umbrella accommodate themselves on the pictorial surface. And, nevertheless, they remain integral, intact. None of the affinities or correlations I have enumerated imply the domination of one form over the other. The things, the transitions, live side by side as equals, free to display themselves, to breathe, to validate themselves, without being assaulted by the 'personality' of their neighbors or by any authoritarian or unifying scheme. The calm aspect of the umbrella, round as a sun, in the foreground, is not troubled by the crushing strength of the metal—although it is not indifferent to it. Conversely, the energy of the metal is not lessened by the impassibility of the umbrella. The allegory could really be an allegory of freedom, of the self determination of a prosperous city."

One sees what degree of sophistication and absurdity such a discourse attains. This is a man who wants to decrypt the *Allegory* at any cost, which is the product of someone who may simply be out to mock the work.

13 We have shown at length in *Métamorphose du bourgeois* that that is precisely what the bourgeoisie wants from art. Enough said.

14 The debate was marked in France by R. Picard's *Nouvelle critique ou nouvelle imposture*, 1965, to which R. Barthes responded with his *Critique et vérité (1965)*, and then S. Doubrovsky in his *Pourquoi la nouvelle critique, Critique et objectivité*, 1966, which described the essential elements of the "feud."

AFTERWORD
The Indefinite Future

Modern art in its complexity, at best, offers an indefinite future. In the middle of the technical world that appears predictable, rigorous, intelligible, and exact, art adopts a counter stance, but it does this in vain. Art in its ambiguity is either the capstone of the closed structure of the technological world or a fissure of uncertainty by which a kind of narrative history could return. But I fear this fissure is only superficial, and may be, in reality, only a deep crack in the decorative façade; nevertheless, the underlying structure remains intact. But, one must try to understand. From the start we see uncompromising stances. The Francastel-Mumford antithesis is well known. Francastel attacks Le Corbusier as a political reactionary and at the same time as a technicist, developing in his capacity as a spokesman for cultural outsiders, theories based on the social myths of the nineteenth century. He also attacks Mumford as an organicist and as a banal humanist, a mystic of progress who takes pride in being subjective and non-scientific. And finally he is against Giedion as the critic of *technique*—which leaves man on the sidelines, mutilated—who cites evidence without proof and who has a narrow view in his appreciation of the degree of evolution in contemporary thought and sensibility. Francastel proceeds exclusively from an enthusiasm for *technique*, from a conviction that socialism will solve all problems, and from the perfectly mythical belief that art will transform the world. He believes blindly that all manner of techniques, including the artistic, are at the service of greater ends; art and *technique* work together harmoniously to produce the new human world by harnessing the arts to the productive activities of society. Man in every case exercises "his demiurgic power." We have already dealt with this ideology in our book, *Technique, the Wager of the Century*. No point in repeating it. But, by contrast Mumford's very measured statement of position in *Technics and Civilization*, then in *Art and Technics*, has become, in turn, very uncompromising in *The Myth of the Machine*, uncompromising because, most likely, it is desperate. He draws up a violent no-holds-barred brief against modern art, which has become for him anti-art,[1] expressing a cult of anti-life. This cult of anti-life, has appeared to him at the same time as anti-matter has appeared to physicists. "Non-art, anti-art, are methods for inducing vast quantities of educated people to relax their already weak hold on reality and to abandon themselves to empty subjectivity … the mark of what is today called authentic experience is the elimination of the good, the true, and of the beautiful … with aggressive attacks against all that is healthy, well balanced, sensible, rational, and motivated. In this world of inverted values, evil becomes the supreme good … a morality turned inside out." "Anti-art professes to be a revolt against our hyper-mechanized, hyper-regimented culture. But at the same time, it justifies the ultimate products of the powers that be: anti-art acclimates modern man to the habitat that megatechnology is in the process of creating: an environment degraded by the discharge of waste—nuclear plants, superhighways, etc.; then, all of this is destined to be architecturally homogenized." "By taking possession of the subjective annihilation that the megamachine portends, the anti-artist gains the illusion of conquering destiny by an act of personal choice. While appearing to challenge the powers that be and to deny the force of routines, anti-art obediently accepts their programmed outcome.

We are, here, in the presence of a stand taken, which is not without foundation and which presupposes a much deeper understanding of technical society, but which relies on a return to certainties about the permanent existence of values, a good, a beautiful, a right way, etc. We can no longer realistically, in this day and age, rely on permanent and universal values (which are the purviews of faith) to evaluate *technique* and Art but must rely, instead, on the measure of man to the extent we have understood the concept of man. If we attempt to leave behind the purely descriptive without falling prey to a judgment based on permanent values, the scientific approach can only come to this. Clearly in the last ten thousand years, there have been periods of crisis, blockage, and turning back, when humans might have thought that everything was being challenged. But I believe there is a great difference between what happened then and what is happening now: the efficiency of our means, the universality of the crisis, the radical nature of controls, the alliance between the material process of degradation and the ideology of intellectuals and elites bring to the fore a qualitative difference between previous crises and our own. Currently we witness a questioning of the entire historic understanding of what it means to be human. Therefore, we must make a choice without being able to say if what is happening today is absolutely good or bad.

The rupture between everything that is done presently under the names painting, sculpture, music, etc., from what these names traditionally suggested is so radical that there is no common measure. The total error of Malraux is to try to understand what is happening now by means of the comparative history of art; he obeys (with more talent than all the others) the desire to embed the present phenomenon in the history of art and to explain it by a discourse that is born of that history. But the misunderstanding is immense. There is in art the same total rupture as in all other activities; *technique* introduces us into a radically new universe, never before seen or thought. Previous knowledge is no longer adequate. One can rightly call this the end of logocentrism; for five hundred thousand years man has been above all a talking animal and all that he has produced has been dictated by logocentrism, and art in particular. Now, abstract painting and concrete music mark the end of this primacy. It is no longer the case of one school opposing another. It is the total rupture with the culture born of logocentrism. Painting and music are dead (as is philosophy!), and we do something else that has nothing to do with the word but something that derives exclusively from means of action. The logos and the word are finished. Now it is the Act (but no longer the personal, heroic act)—but the mechanical act. Hence, we either try to defend a slow development of the Good, the Beautiful, the Human or we erase it all with a stroke of the pen and start again from zero—that is the issue. In the domain of music we have believed in a kind of serious music, a classical music—but this was an historic moment. In the age of the transistor and the hifi, in the continual flood of sound, we cannot recall the real experience of music. Instead of the true music lover, we have the casual listener (connoisseur) as expert. This approach to music—musical technique—kills musical emotion. The pretentious connoisseur kills true knowledge. The casual connoisseur analyzes the most learned counterpoint but no longer has a complete sense of what the piece of music is about or any sense of its worth. The best of modern art is the unsayable, the inexpressible expression of human suffering, the crucifixion by the inhumanity of technique, the suffering that art does not transfigure nor symbolize but rather expresses in its raw nakedness that no longer allows play or distance; instead, suffering is thrown in the face of the listener or spectator. Furthermore, suffering is not translated by the listener but is experienced directly with the enormous volume of pop music and with the nervous exasperation produced by Xenakis or Barbaud; *anguish* is produced in the spectator as an angst that becomes the sole, common measure and means of communication for present society. This anguish is produced to the degree that art upholds a negative ahistoricity and a rigid hierarchy. Thus art reinforces and confirms the reality of the technological society: far from being a protest or a taking of account, art shuts the door by legitimating for the listener the technical reality that nevertheless is alien to him. It makes illusion out of reality and gives reality to the illusory. It prevents man from understanding the technological world by immersing him in it, and, meanwhile, by focusing him on appearance and insignificance. Such is, among other things, the role of art with a political message. The real anguish experienced, because of technical conflict, is diverted toward political activity, but the means of aesthetic diversion embeds man more deeply in the technological system by reproducing more powerfully the motivations of his anguish.

To eliminate meaning from art is, in reality, to eliminate "what man has lived for until now." It is effectively to eliminate man. Undoubtedly, there is a kind of "outdated romanticism." "Let us delve into the deep structures of immediate sensitivity." But this is precisely what man has tried to escape! The relativization and the reduction of the human in morality and aesthetic canons have led to: "what we construct artificially and arbitrarily." There is no longer a face, or hair, or trees, or clouds: all that was conventional: now there are lines, colors, elements of perception, and a certain arbitrary order in combining them. As a result, one confuses the arbitrary nature of rules for playing bridge with those that allow man to construct a reality on his own terms which he would truly live, experience, and master. Modern art, as we have often said, has killed the subject in the double meaning of "theme," chosen for the work and also in the sense of an active subject, composer, executor, active listener, only to leave no more than a mechanism for gratuitous abstraction. So be it. But, one must see what this produces: one can argue that an absolutely pure artistic creation is produced. This is evident. An absolutely artificial scheme leading to the creation of a completely fictitious universe permits absolute purity. But, man has never been able to live in any kind of total purity. Only an algebraic sign would succeed at that. In this purity, the object is the only possible subject of interest. What counts is the technique of production, the process of communication, and the method of creation. This clearly implies the elimination of the subject and lived time in favor of an exclusively mechanical time. Such is the universe of presuppositions but also of expected consequences if one achieves this aesthetic technological purity. Now, the destruction of the subject and of time leads to, among other things, a deep-rooted conformity. It is very interesting to note that artists (in all areas of endeavor) claim to be non-conformists, whereas all the presuppositions of modern art lead inevitably to the conformity of listener or of spectator. But this does not lead to political or economic conformity but rather to the deeper, more thoroughgoing technical system and also to the idea of majority rule: what certain theoreticians of art call the present social consensus and offer as the sole criterion of art. Either there is an "eternal" value of beauty, of the good, etc. against which one measures art, or else, there is group consensus, which one must obey: pleasing them and producing for them is the only possibility. And conformity in art will be all the more rigorous as art throws off the shackles and vagaries of sense to achieve efficiency of impact. Technology not only impinges on the means of production but has infiltrated the artistic act by subordinating it to the imperative of calculated efficiency. From instinct and intuition the painter once felt that placing a brush stroke would produce a certain effect or impression; and the musician felt that a certain chord would induce a certain emotion, but he knew this simply because he felt that emotion, because he was seized or overcome by that tonality, which produced a fascinating shock. We have changed all that. Now, thanks to the techniques of psychology and communication, the spectator can be completely possessed, snared in an inescapable sensorial net, bewitched by a sensual deluge: according to Moles, "therein lies a brand new role for the computer in art, which has not been fully exploited." To be sure, this tack has been taken, and because of this, one can say that today we no longer know what art is and to what it answers. *Its field of operations has changed.* But, it is impossible to make art a simple game, unless you are willing to call the scientific activity leading to the atomic bomb a game. Art has become one of the primary forces of integrating man into the technological complex, and that is why art is, as well, the bearer of a political-philosophical message of pure entertainment, perfectly bereft of significance and meaning, because man cannot abide with all the dimensions of the technological reality.

The return to the raw element—sound as root datum, the graphic sign, the stroke of color, the word become sound—abstracts from the cultural relations the meanings attributed to these elements and becomes an integral part of functionalism.[2] This attitude, as we have seen, is often considered a sign of progress. "Musicians have the audacity and honesty to break with all conventions to return to the basis and origin of music: the raw block of sound." But, in reality, this operation unhinges man from all he found meaningful and throws him into the unknown realm of the technological system. He is denied references to the past; he can no longer judge what is occurring based on the moral, aesthetic, or ideological grounds, which allowed him to maintain mastery *vis-à-vis* the technological explosion. On the contrary, he must, under the guise of breaking with outmoded cultural patterns, return to an undeniable base and shatter all traditional cultural landmarks and points of reference. But, those who undertake this task—the modern artists—so praised for their audacity, while being essentially conformists seem never to reflect on some basic questions: "Why did the so-called primitive man, hearing pure noise, raw sound, experience the need to elaborate and to construct something other than noise and sound? Why did he match sound to meaning? Why was he never content simply to repeat raw sound?" And, moreover: "Is it the point of art to renounce interpretation, elaboration, and symbolization?" Here is where modern aesthetic theory leaves us!

Specifically, the cleverly constructed attack on language is the leading edge of integration into the technological system. The lie behind the revolutionary message of art springs from the fact that art limits itself to consecrating the doctrine of anything goes. Or, on the other side, we have the Lacanian position that language is completely shattered and conveys nothing, because there is nothing to convey other than the spoken word, turned inside out, tortured, and disintegrated. There is nothing but "structures" that one can arrange and construct like Lego blocks. Language has become a Lego set, and it is clear that it is now treated as such. If the constructor reveals something in his structures, it is not what he intended to say but rather something else that requires a technical interpretation that strips man of himself by revealing what was hidden or else by forcing him to accept truth guaranteed by the technician. It is, therefore, necessary to have blind confidence in the technician or psychoanalyst who claims to see what no one else knows. The truth is that, if man comes to question art and language on the basis of *technique* or its impact, it is because he has been brought into question by *technique*. To the extent that his being is on the edge of nonexistence, an aesthetic vacuum and an absence of meaning dominate. The very being of man is bled out as science and mechanisms rule his life. Thus, man explains his own nonexistence (and this is one of the essential roles of art with a message) and reestablishes himself as a challenger of traditions. Art is the challenger's vehicle that suggests a new and autonomous activity that recalls art's prior elevated status. Now, the only function of art is to challenge precisely because technology questions the being of man. Now, the calling to question of all meaning is mirrored (and it could not be otherwise) in a perfect conformity with action and behavior. Conformity is necessary for the technological system to function in its highest sense with the reproduction of revolutionary attitudes, which square off against the nonexistent powers of the past. The technological process attacks the revolutionary on a front that no longer exists. Technology, as well, questions man's very being (something that he deliberately accepts) and imposes total conformity to unacknowledged and unspecified forces that structure this new environment. Under these conditions, modern art is both a witness and an accomplice. It witnesses man's abandonment of self-knowledge, his renunciation of understanding (other than by development of scientific paradigms), and his refusal to organize his world (apart from technical structures). Art witnesses the omnipresence of the technical system in as much as the domain of the senses has become a play of technical structures. Art witnesses the existential void. If one continuously proclaims, "There is nothing to say," it is because there is no longer anything to *experience*, and human science is the dangerous proof of this fact. We no longer have novels that tell the story of a hero, because the hero of modern times no longer experiences anything and does not have a story. He exists through what happens to him like the suited cosmonaut.

If one shows in art that man is an object, that the artist himself is an object, that one can disfigure this absence of the real, we only see the object replacing man (on the one hand) and the way *technique* treats man (on the other hand). In other words, we have a recognition of the absence of that man to whom much excessive importance has been attributed (Robbe-Grillet). But this is not a total critical absence: in opposition to the absence of man, unacknowledged but legitimating forms and structures subsist alongside *technique* itself. Thus, art becomes, and this is its other face, the accomplice and assumes a falsifying role (which it does not necessarily have in all societies). It is an accomplice to the annihilation, to the nonexistence, and then to the reification of man: it does not allow man to live well but, on the contrary, drives him deeper into despair. Art justifies the status quo (the triumph of *technique*) and offers a few compensations for his intolerable situation (television and cinema); it offers the illusion of revolt, of taking the initiative, of freedom, but each time the products of this illusion further nourish the conviction that man is nothing but a cosmic accident (a few daubs of yeast appearing by accident) and that he cannot attribute meaning or value to anything whatever. This formless and obsolete art only confirms his nonexistence. Even those who do not go to see painting exhibitions, those who do not go to hear serial music, are impregnated by the defeatism broadcast by the mass media through popular entertainment. Art becomes an accomplice by inculcating what *technique* has made of man, by preventing him from protesting too much, by producing idealist justifications for reality, and by dressing conformity in the appearances of freedom. Art is also the direct expression of the dominant factor, which is not ideology and not the preeminence of the bourgeoisie. *Technique* produces, thus, a totalizing ideology, from which man can no longer be distant, simply because this process of symbolization and distancing, which was once a specific and integral part of art, is now completely devalued and rendered meaningless, watered down, and even ridiculous by technological activity, which does not symbolize and is only efficacious.[3] We are integrated into the technological system that exploits symbolization for its own ends and development and produces the distancing that is necessary for *technique* to progress by incorporating everything into the system. Thus, according to this description and analysis, modern art is in no way—in none of its dimensions or expressions—a creator, a liberator, or a means of freedom. As things stand, there is no exit, either for art or for man.

Is there no hope? At first sight, no. But there is always hope. But this hope implies a series of decisions and choices. If we continue to blindly recite that "something speaks," that, "man is an accident," that "chance and necessity" are everything, that "everything is political," that "everything comes down to class struggle," that "art must be modern," that "everything is relative," etc., then, in effect, we can expect and do nothing. Art can only rediscover its critical force and voice if it abruptly breaks off with the technological system, if it ceases to function in a raw and permutational mode, if it ceases to obsess with new materials and mechanisms; only with these cessations can art return to values, ethics, and meaning. This is the choice that we face. Of course, this does not mean a repetition of traditional values, a return to bygone meanings and bourgeois ethics! No, art must above all be inventive (beyond being modern), but inventive *in a true sense*, and nothing else. The rest is folly. Art must reaffirm meaning (and not get lost in nonsense) and be *against* nonsense, and as such, it must be the locus of rupture, of challenge, of an effective accusation of the technological system. The artist must not remain a simple manometer to measure the pressure of meaning, a simple seismograph to register the temblors caused by successive waves of technique. To the degree that art is "modern," and to the degree that it reflects "nonsense" and the absurd, it will be nothing more than the jumping jack that distracts the onlooker as his pocket is picked. And I say this about the greatest of them all—Xenakis as well as Paul Klee, Butor as well as Albee, Beckett as well as Vasarely, Stockhausen as well as Giacometti, the creators of spaces, the entrepreneurs of sound, the eulogists of objects. No one can follow the rapidity of modern techniques and mechanisms. No one who enters this game can affirm the superiority of man. What is unique: precisely at the moment when everyone agrees on the need to redirect and manage the trajectory of technological development, those who would be the most suitable to undertake this task—philosophers, theologians, artists, declare themselves out of the game and would rather announce the end of man. Further, we have the need to retrieve a sense of meaning—one that is not hopeless or suicidal—and one that allows life among the demons; a sense of meaning that offers both significance to our life and direction for our will. And, finally, we acknowledge the need to symbolize reality in a way that is not blind fate but that, because it is dominated by meaning, enables man to persist amid this rational chaos. Is this a hopeless cause? On the contrary, I believe that modern man deeply needs meaning and calls for it with so much energy that the outcome is assured. But, this still must be a meaning that liberates itself from politics for otherwise only dictatorship and oppression will follow. I know to what point this will scandalize some, but so long as one seeks meaning in the political path, so long as art attempts to say something by producing propaganda, one will keep on submitting to the curse of our time. The political path is radically blocked (and let it rest in peace. Definitely! Of course, it's our only way to escape politics). In the political realm one cannot escape the double play of its essential nature between power and communication to which modern art perfectly panders. Power, which tends toward the absolute (and it must do this when there is neither sense nor value), increasingly operates in secret, communicates less and less, and proceeds with abstract and incomprehensible paradigms for the non-technician; but, at the same time, thanks to the mass media and the proliferation of communication, false meaning and false values are diffused, which have nothing to do with any reality. Power diffuses what it does not believe, does not do, will not do, by hoodwinking the public, who is pulled into a game of pure illusion to the extent that they believe the saturating mass media.[4]

By necessity, the politician manipulates a double discourse: one discourse is objective and meant for the inner circle of party members to reveal what he will truly do at the level of pure praxis and a Machiavellian strategy to attain ends; and the other discourse, directed toward the public, is clothed in absolutes aiming for a groundless reality promulgated by Art and the mass media, of course! Only a radical challenge to the political system will enable the reestablishment of meaning and value beyond those tricked out by propaganda. But this challenge to politics does not imply a retreat to the "ivory tower": to the contrary, symbolizing the reality of today puts in motion a force that questions the reality of tomorrow, which would lay the ground work for another type of society. This was the role of creative activity in art each time it fulfilled its plenary function that is the fundamental recognition of what is (the Janus face of State and *technique*) in the name of a new birth. Art is the procreator, first, and then the midwife. But, to that end, it is necessary to shake off the general defeatism of all modern artistic activity for the past century. Art would now assume the task of rendering our technical universe intelligible and manageable and of encouraging man to rise up against servitude and progress tied to consumerism, to stir up the ever-present and burning desire for the absent. Wholly Other. We have a desire experienced as an absence opposed to the laws of necessity and adaptation that embrace the view that happiness is just around the corner. First and foremost, art is a testament to the combination of desire and absence. But art would have the force to establish desire not as a morbid embrace of absence but as a movement toward the creation of something new. The task is ahead. We must realize that art for the last century has taken a totally false road. It has lowered itself to the powers that be and has born witness only to the defeat of man. We must also recognize, then, that art does not proceed by way of science to rediscover its value and its truth. Meaning and value do not come from outside or from intellectuals, philosophers, or moralists. The artist must be again the creator, not merely of forms and combinations, but must instead be the progenitor for the group and society whose meaning and values he embodies. And he will accomplish this not through an empty intellectual or moralizing discourse but through a profound rediscovery, which informs his entire work, making it available and understandable.[5] Heinrich Böll, Nathanael West, Marc Chagall, Salvador Dalí, Luis Buñuel, Ingmar Bergman, Federico Fellini, Saint-John Perse (Alexis Saint-Léger Léger), Pablo Neruda, all have failed. Who will set sail on the high seas?

What is needed is an essential understanding that is not merely existential or phenomenological. What is needed is courage to defy the computerized demons. What is needed is confidence to advance and to conquer. These are three columns on which, perhaps, art can build and which would provide the exodus from this time and the exordium of a new discourse that we must have to escape the mathematics of destiny.

Notes

1 Indeed Mumford means by anti-art the *totality* of contemporary art since 1945.

2 When I write this, I'm not referring back—let us clarify—to the general theory of sociological functionalism. But I simply note in our society a de facto functionalism of essential elements, art among others, which are now validated only in relation to their function within the system.

3 Art as a whole falls under the spell of the simplistic story so often recounted in the anti-religious propaganda of the early stage of the Russian Revolution: they had children plant two gardens. In one they would place seeds and fertilizer, and then water and tend the garden every day. In the other, nothing. But they would say prayers and have a priest's blessings. The first produced vegetables; the other—nothing. This idiotic tale is in reality the dominant story of our society, which is retold every day in a million ways. This is, for example, used as evidence for the superiority over discourse and of the audio-visual (but for teaching what?) over rhetoric. Art is summoned to enter the first of these paths or to be nothing, and above all to be *accepted for nothing.*"

4 For more information on this double play that demonstrates current political reality, see my *The Political Illusion*.

5 "All this depends on imagining human possibility, at least as understood, as complex and as surprising, as contradictory as the atomic possibility. What is lacking in the approach to man is both respect and meaning. By respect, I understand the care that one will take in an experimental approach as has been done with the atom and which will not be limited to old wives' tales or to electrodes in the brain. By meaning, I understand that the human universe is to be explored in the contrary direction. Concerning the cosmos whose mechanism is beyond us, we have only to take apart the timepiece. Humankind, on the contrary, is to be identified and to be formed. We are still with detached pieces: individual and groups, warring tribes, primitive confrontations. With the stubbornness of an idiot, man turns away from these forces of rupture and cohesion, which he barely understands and which he refuses to recognize. This sluggard who flails about would like to learn the lesson in the clouds by looking through the window. He would like to copy the closed systems of nature; he seeks the summations of a calculator. Nothing says the system of man is so closed or confined. Nothing says that we have understood man's essential nature. Nothing says that his dominant quality is intelligence, at least as we know it, and of which we have made in our history such poor use." (P. Schaeffer).

BIOGRAPHICAL NOTES

Jacques Ellul was a French sociologist, historian of legal institutions, lay minister, philosopher, and educator. He studied literature, history, politics, and law, writing his dissertation on Roman Law in 1939. In the 1930s, he joined the fight against fascism, and in 1940 he openly opposed the Vichy government. Later, he fought in the *Résistance*, and received the *Légion d'honneur*. From 1947 onward, he taught history, sociology and law at the Université de Bordeaux, where he served briefly as deputy mayor. He was at once a Protestant Christian as well as a Marxist thinker, refusing to abandon either one; and developed a theology which was strongly marked by Barth and Kierkegaard. From the mid-1940s, Ellul began to develop his theory of *technique*, which he saw as the usurpation of all the symbolic and material construction of the modern age by utmost rationality and efficiency. He painstakingly analyzed *technique*'s pervasive influence as an order that frames the contemporary mind in all fields of endeavor in a long list of publications, the three most important being: *La technique ou l'enjeu du siècle* [The Technological Society] (1954), *Le système technicien* [The Technological System] (1977), and *Le bluff technologique* [The Technological Bluff] (1988). In his long scholarly career, Ellul wrote forty books and eight hundred essays. He received the *Prix d'histoire* from the *Académie française* for his massive study of the history of institutions: *L'Histoire des institutions* (1955). Other publications of note are: *Le vouloir et le faire* (1964); *L'illusion politique* [Political Illusion] (1965); *Sans feu ni lieu* [The Meaning of the City] (1970); *Ethique de la liberté* [The Ethics of Freedom] (1973-75); *L'empire du non-sens* (1980); *La parole humiliée* [The Humiliation of the Word] (1981).

Samir Younés is Professor of Architecture at the University of Notre Dame, USA. His latest book is *The Imperfect City: On Architectural Judgment*, 2013.

David Lovekin is Emeritus Professor of Philosophy at Hastings College, USA. He is the author of *Technique, Discourse, and Consciousness: An Introduction to the Philosophy of Jacques Ellul*, 1991.

Michael Johnson is Emeritus Professor of Foreign Languages at Hastings College, USA.